Laboratory Manual
Teacher Guide

HOLT, RINEHART AND WINSTON

A Harcourt Education Company

Austin • Orlando • Chicago • New York • Toronto • London • San Diego

Holt Science Spectrum
Laboratory Manual
Teacher Guide

Cover: Pete Saloutos/Corbis Stock Market

Unless otherwise noted all other illustrations by Holt, Rinehart and Winston.

Page T47 (Fig. A), Thomas Gagliano; T65–67 (Fig. 1-4), Thomas Gagliano; T92 (Fig. 1), Mark Heine.

Printed in the United States of America

ISBN 0-03-067077-2

7   8   9   10 11 12   082   10 09 08 07 06

Teacher Guide Contents

Teacher's Introduction

Student's Introduction

Lab 1 Introduction to Science

Lab 2 Matter

Lab 3 Atoms and the Periodic Table

Lab 4 The Structure of Matter

Lab 5 Chemical Reactions

Lab 6 Solutions, Acids, and Bases

Lab 7 Nuclear Changes

Lab Skills and Scientific Methods

SKILLS PRACTICE LABS DEVELOP PROCESS SKILLS

The twenty structured experiments in this book provide more in-depth investigations of chapter topics. They also provide additional opportunities for your students to develop crucial laboratory process skills.

The "Skills Acquired" box appears in the Teacher's Notes and Answers for each lab. The information in this box will provide you with a list of specific skills that each student should acquire by performing a lab activity. The following is a list of process skills that a student may acquire by performing the labs in this book.

- Classifying
- Collecting Data
- Communicating
- Constructing Models
- Designing Experiments
- Experimenting
- Identifying and Recognizing Patterns
- Inferring
- Interpreting
- Measuring
- Organizing and Analyzing Data
- Predicting

STUDENTS GAIN EXPERIENCE USING THE SCIENTIFIC METHOD

Although there are many methods used in science, the following steps describe a method that can be used to guide students' scientific inquiry. As students work through the activities in this book, they will gain experience using each of these steps and working through the scientific method to solve problems.

The Teacher's Notes and Answers for each lab include a list of the steps of the scientific method used in that lab. This list will also highlight the point at which the student uses that step of the scientific method during the lab. The following list gives the steps of the scientific method as developed in this program. Students will follow each of these steps as they work through the scientific method, and some activities will focus on one or more of these steps.

- Make Observations
- Ask Questions
- Test the Hypothesis
- Analyze the Results
- Draw Conclusions
- Communicate the Results

Teacher's Notes and Answers

Comparing the Densities of Pennies

TIME REQUIRED
1 lab period

SKILLS ACQUIRED
Classifying
Collecting data
Experimenting
Recognizing patterns
Inferring
Measuring
Organizing and analyzing data

THE SCIENTIFIC METHOD

• **Make Observations:** In the Procedure, students sort pennies by date and measure masses and volumes.

• **Analyze the Results:** In the Analysis, students calculate the average densities of two groups of pennies based on their measurements.

• **Draw Conclusions:** In the Conclusions, students draw conclusions about how the composition of pennies changed in 1982.

• **Communicate Results:** In the Analysis and Conclusions, students communicate results by providing written answers to the questions.

Teacher's Notes
MATERIALS

Be sure that each group of students has both pre-1982 and post-1982 pennies. The numbers of the two types of pennies do not have to be equal, but having at least three or four pennies in each category will make the volume easy to measure. To avoid confusion, do not give students pennies made in 1982.

SAFETY CAUTIONS

Remind students to be careful when working in the lab. Warn them to be careful with the graduated cylinders if the graduated cylinders are made of glass.

TECHNIQUES TO DEMONSTRATE

Demonstrate proper techniques for using the balance. Also demonstrate how to properly measure volume using the meniscus in a graduated cylinder.

DISPOSAL INFORMATION

Students should dry the pennies and place the graduated cylinders on drying racks when they are finished with the experiment. Instruct students to return the pennies to you after the lab, and be sure to keep track of how many pennies each group has.

Name _____ Class _____ Date _____

Skills Practice Lab

Comparing the Densities of Pennies

Introduction

All pennies are the same, right? All pennies are about the same size and color—and all are worth one cent—but not all pennies have exactly the same physical properties. Individual pennies may differ from one another based on how much they have corroded or how much dirt they have accumulated. A more significant difference may be in the composition of the metals that make up the pennies. Occasionally, the U.S. Department of the Treasury changes the composition of pennies, which changes the overall mass and density of each coin. In this experiment, you will determine the average density of two group of pennies made at different times. Then, you will compare the average densities of the pennies in the two groups.

OBJECTIVES

Construct the average densities of pennies by using mass and volume measurements.

Compare the average densities of pennies made before 1982 and after 1982.

Infer the change made in the composition of pennies in 1982.

MATERIALS

balance   paper towels
graduated cylinder, 100 mL   pennies (10)
paper   water

Safety

- Secure loose clothing and remove dangling jewelry. Don't wear open-toed shoes or sandals in the lab.

- Check the condition of glassware before and after using it. Inform your teacher of any broken, chipped, or cracked glassware because it should not be used.

- Do not pick up broken glass with your bare hands. Place broken glass in a specially designated disposal container.

Procedure

1. Sort your pennies into two piles. One pile should contain pennies made before 1982. The other pile should contain pennies made after 1982. If you have a penny made in 1982, ask your teacher to replace it with a different penny. Record the number of coins in each pile in a data table like the one shown on the next page.

Name _____ Class _____ Date _____

Comparing the Densities of Pennies (cont.)

2. Use the balance to measure the combined mass of the pennies in the pre-1982 group. Record the mass in grams in your data table. Then, use the balance to measure the combined mass of the pennies in the post-1982 group. Record the mass in grams in your data table.

3. Fill a graduated cylinder about halfway with water. Note that the top surface of the water curves upward toward the edges of the cylinder. The curved surface of the water is called a *meniscus*. You should always use the bottom of the meniscus when measuring volume in a graduated cylinder, and you should always put your eye at the same level as the meniscus when taking your volume reading.

4. Measure the volume of water in the graduated cylinder by using the bottom of the meniscus. Record the volume in milliliters in your data table.

5. Now add the group of pre-1982 pennies to the water in the graduated cylinder. Measure the new volume of the water and pennies combined. Record the volume in milliliters in your data table.

6. Carefully remove the pennies from the graduated cylinder, and pour the water into a sink. Use a paper towel to dry the pennies.

7. Repeat steps 3–6 for the group of post-1982 pennies.

8. When you are finished, place the graduated cylinder on a rack to dry and return the pennies to your teacher.

DATA TABLE

Group	# of coins	Volume of mass (g)	Volume of water (mL)	Average water and coins (mL)	Coins (mL)	Density of coins (g/mL)
Pre-1982						
Post-1982						

ANALYSIS

1. **Organizing data** Calculate the volume of the pre-1982 pennies by subtracting the volume of the water from the combined volume of the water and coins. Record this volume in your data table. Repeat the calculation for the post-1982 pennies, and record the result.

Answers may vary depending on the data.

_____

_____

_____

Name _____ Class _____ Date _____

Comparing the Densities of Pennies (cont.)

2. Organizing data Calculate the average density of the pre-1982 pennies by dividing the mass of the coins by the volume of the coins and then dividing the result by the number of coins in the group. Record the average density in your data table. Repeat the calculation for the post-1982 pennies, and record the result.

Answers may vary. The average density of pre-1982 pennies is about

8.87 g/mL. The average density of post-1982 pennies is about 7.18 g/mL. If

student answers are way off, make sure that students divided by the number

of pennies in each group to get an average.

3. Analyzing results Which group of pennies has a greater average density?

The pre-1982 group of pennies has a greater average density.

_____

_____

_____

CONCLUSIONS

1. Drawing conclusions Why do you think that pennies made before 1982 have a different density than pennies made after 1982 do?

The U.S. Treasury changed the composition of pennies in 1982. (Other

answers are possible. Accept all reasonable inferences.)

_____

_____

2. Applying conclusions Pennies made before 1982 and pennies made after 1982 contain a combination of copper and zinc. The density of copper is greater than the density of zinc. Is the ratio of copper to zinc in the pennies made after 1982 higher or lower than the ratio of copper to zinc in the pennies made before 1982?

The ratio of copper to zinc is lower in post-1982 pennies than in pre-1982

pennies. The average density is lower in post-1982 pennies because the

composition of the pennies changed.

_____

Name _____ Class _____ Date _____

Comparing the Densities of Pennies (cont.)

EXTENSIONS

1. **Research and communications** Research changes in the composition of pennies made by the U.S. Department of the Treasury in the past century. Write a paragraph describing the changes and explaining why you think the changes were made.

The other major change to the composition of pennies took place in 1943. In that year, pennies were made of steel plated with zinc. This change was necessary because of a shortage of copper during World War II.

Teacher's Notes and Answers

Measuring Density with a Hydrometer

TIME REQUIRED
1 lab period

SKILLS ACQUIRED
Communicating
Constructing models
Experimenting
Interpreting
Organizing and analyzing data
Predicting

THE SCIENTIFIC METHOD

- **Make Observations:** In the Procedure, students observe how a hydrometer behaves in different liquids.

- **Form a Hypothesis:** In steps 7 and 9 of the Procedure, students form hypotheses about the relative densities of liquids and the resulting behavior of the hydrometer.

- **Analyze the Results:** In the Analysis, students organize data and describe the events in the experiment. They also explain the results in terms of buoyancy.

- **Draw Conclusions:** In the Conclusions, students draw conclusions about the relative densities of the various liquids tested in the experiment. Students also compare their results to their predictions and offer explanations for the differences in density by using the concepts of mass and volume.

- **Communicate the Results:** In the Analysis and Conclusions, students communicate their results by providing written answers to the questions.

Teacher's Notes
MATERIALS

The regular and diet cola that you provide should be made by the same company (to eliminate other differences in composition). Both colas should be fresh, or both should be flat.

SAFETY CAUTIONS

Remind students not to drink any of the liquids, even colas from cans in the lab. It is not safe to eat or drink in the lab.

DISPOSAL INFORMATION

All liquids may be disposed of in a sink. Plastic straws may be discarded in a trash can. You may want to dry and reuse the modeling clay, or it can be discarded in a trash can.

Measuring Density with a Hydrometer (cont.)

TIPS AND TRICKS

After the experiment, you may wish to perform a demonstration to test student predictions in question 3 of the Conclusions. Drop an unopened can of diet cola and an unopened can of regular cola into a bucket or sink full of water. The can of diet cola will float, and the can of regular cola will sink.

Name _____ Class _____ Date _____

Skills Practice Lab

Measuring Density with a Hydrometer

Introduction

A hydrometer is a device used for measuring the density of a liquid. Hydrometers have a wide variety of uses. For example, beverage makers use a hydrometer to test their product at different stages during production. The density of the beverage serves to indicate whether certain chemical processes have taken place in the production process. An ecologist can use a hydrometer to test the salinity of water. This test tells the ecologist whether the water is a suitable habitat for different kinds of aquatic organisms. In this experiment, you will construct a simple hydrometer. You will not use your hydrometer to measure exact values. Instead, you will use the hydrometer to compare the densities of fresh water and salt water and to compare the densities of regular cola and diet cola.

OBJECTIVES

Construct reaction time of a falling object by using falling-distance data.

Compare reaction times measured by using a stopwatch with reaction times calculated from falling-distance data.

Explain the methods used to determine reaction time in the experiment.

MATERIALS

balance	paper towels
beakers, large (4)	plastic drinking straws
cola, both regular and diet	salt
colored permanent markers	scissors
masking tape	stirring rod
metal BBs (6)	water
modeling clay	

SAFETY

- Secure loose clothing and remove dangling jewelry. Don't wear open-toed shoes or sandals in the lab.

- Do not pick up broken glass with your bare hands. Place broken glass in a specially designated disposal container.

- Check the condition of glassware before and after using it. Inform your teacher of any broken, chipped, or cracked glassware because it should not be used.

- Use knives and other sharp instruments with extreme care. Never cut objects while holding them in your hands. Place objects on a suitable work surface for cutting.

Name _____ Class _____ Date _____

| **Measuring Density with a Hydrometer (cont.)**

Procedure

1. Fill a beaker about half full with water. Label this beaker "Fresh water" by using the marker and a piece of masking tape.

2. Fill another beaker about half full with water. Pour about 4 Tbsp of salt into the beaker, and stir with a stirring rod. Label this beaker "Salt water."

3. Fill another beaker about half full with regular cola, and label the beaker "Regular cola." Fill a fourth beaker with diet cola, and label the beaker "Diet cola." Caution: Do not drink any of the cola or eat or drink anything else in the lab.

4. Cut a plastic drinking straw in half. Seal one end of the straw with modeling clay. Hold the straw vertically so that the open end of the straw points up. Drop two BBs into the open end of the straw. You now have a simple hydrometer.

5. Place the hydrometer, clay end first, into the beaker of fresh water. Adjust the hydrometer by adding or removing BBs until the water level is about halfway up the body of the hydrometer.

6. Use a marker to mark the water level on the hydrometer when the hydrometer is in fresh water. Pin the hydrometer against the side of the beaker, and hold the hydrometer in place with your fingers).

7. Predict whether the salt water is denser or less dense than the fresh water? Will this difference in density cause the hydrometer to sink more or less into the salt water than into the fresh water? Discuss your prediction with the other members in your lab group.

8. Remove the hydrometer from the fresh water. Place the hydrometer, clay end down, into the beaker of salt water. Mark the water level on the hydrometer by using a marker of a different color than the marker you used for the fresh water line. (Tip: You can mark your beaker labels with the colored markers so that you can remember which color corresponds to which liquid.)

9. Predict whether regular cola or diet cola is denser. In which liquid will the hydrometer sink the least?

10. Repeat step 8 for the beakers of regular cola and diet cola. Use a different colored marker to mark the water level in each beaker.

ANALYSIS

1. **Organizing data** List in order from bottom to top the colors of the lines on the hydrometer. Then repeat the list, substituting the names of the corresponding liquids for the colors.

   Answers may vary depending on the colors used. The order of liquids will

   most likely be diet cola, fresh water, regular cola, and salt water.

Name _____ Class _____ Date _____

Measuring Density with a Hydrometer (cont.)

2. **Describing events** Did the hydrometer sink lower in the fresh water or in the salt water? How does this result compare with the prediction you made in step 7 of the Procedure?

The hydrometer sank lower in the fresh water. Comparisons will depend on

predictions.

3. **Describing events** Did the hydrometer sink lower in the regular cola or in the diet cola? How does this result compare with the prediction you made in step 9 of the Procedure?

The hydrometer sank lower in the diet cola. Comparisons will depend on

predictions.

4. **Explaining events** Buoyancy is the force with which a liquid pushes upward on a less dense substance. Buoyancy increases as the density of the liquid increases. If a hydrometer is moved from one liquid to a second liquid that has greater density than the first liquid, would the hydrometer sink into the second liquid more or less than it sank into the first liquid?

The hydrometer would sink less into a liquid that has a greater density.

CONCLUSIONS

1. **Drawing conclusions** Which is denser: fresh water or salt water? Explain why. Use the concepts of mass and volume in your explanation. Compare your answer with your prediction in step 7 of the Procedure.

Salt water is denser. When salt dissolves in water, the mass of the solution

increases, but the volume stays the same. As a result, the density increases.

Comparisons will depend on predictions.

Name _____ Class _____ Date _____

| *Measuring Density with a Hydrometer (cont.)*

2. **Drawing conclusions** Which is denser: regular cola or diet cola? Form a hypothesis to explain your answer. Use the concepts of mass and volume in your explanation. Compare your answer with your prediction in step 9 of the Procedure.

Regular cola is denser. Accept all reasonable explanations. Sample answer:

Diet cola contains high-potency sweeteners, while regular cola contains

sugar. Less sweetener is required in the diet cola, so the overall mass is less

while the volume of the cola is the same. As a result, the density is lower.

Comparisons will depend on predictions.

3. **Applying conclusions** If you placed a can of diet cola and a can of regular cola in a bucket of fresh water, what would happen to the cans? Explain your answer.

Accept all answers that have reasonable explanations. Sample answer: The

diet cola would float, and the regular cola would sink. The density of diet

cola is less than the density of fresh water, but the density of regular cola is

greater than the density of fresh water.

EXTENSIONS

1. **Designing experiments** Design an experiment to compare the density of fresh cola with the density of flat cola. Predict which cola will have a greater density.

Evaluate answers based on the quality and detail of experimental designs. Students could reasonably predict that the density of flat cola will be greater than the density of fresh cola. (The fresh cola contains carbon dioxide gas, whose density is very low relative to the densities of the rest of the ingredients; as a result, the overall density of the cola is decreased.)

Teacher's Notes and Answers

Drawing Atomic Models

TIME REQUIRED
1 lab period

SKILLS ACQUIRED
Classifying
Constructing models
Recognizing patterns
Inferring
Organizing and analyzing data

THE SCIENTIFIC METHOD

- **Make Observations:** In steps 1–3 of the Procedure, students observe an atomic model and learn to interpret the model.

- **Analyze the Results:** In the Analysis, students recognize patterns in the results in their data tables.

- **Draw Conclusions:** In the Conclusions, students draw conclusions about the relationship of the features in their atomic models to the properties of the atoms and the position of the atoms on a periodic table.

- **Communicate Results:** In the Analysis and Conclusions, students communicate their results by providing written answers to the questions.

Teacher's Notes
MATERIALS

This activity requires paper only. Students do not need to be in a lab room to complete this activity. Make sure that students have access to a complete periodic table, either on a wall chart or in their textbooks.

Name _____ Class _____ Date _____

Drawing Atomic Models

Introduction

Look at the many different things in your classroom or lab: desks, chairs, windows, laboratory equipment, students, shoes, and notebooks. If all of these things are made from atoms and all atoms are made of only a few kinds of particles, what accounts for the variety of things that you see?

Atoms of different elements have different numbers of protons in their nuclei. In atoms that have a neutral charge, the number of electrons around the nucleus equals the number of protons in the nucleus. As it turns out, the number of electrons in an atom is one of the most important things in determining the chemical properties of the atom.

Elements are arranged in the periodic table according to the number of protons in each atom, which also corresponds to the number of electrons in the atom. The periodic table is called "periodic" because as the number of protons increases, certain chemical properties appear over and over again—periodically. In this activity, you will draw models of atoms and will place the atoms in the proper place on a periodic table.

OBJECTIVES

Draw models of atoms that show the numbers of protons and neutrons and electrons in proper energy levels.

Locate the proper position of atoms on a periodic table.

Infer the relationship of the number of energy levels and number of valence electrons in an atom to the group and period of the atom on a periodic table.

MATERIALS

paper   periodic table of the elements
pencil

Procedure

DRAWING MODELS OF ATOMS

1. **Figure 1** shows a model of an atom. This type of model is sometimes called a *Bohr model* because it was first used by the physicist Niels Bohr. The model shows a nucleus in the center that has three protons (p^+) and four neutrons (n). Surrounding the nucleus are three electrons (e^-). Notice that the number of electrons equals the number of protons in the nucleus.

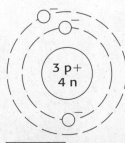

FIGURE 1

Name _____   Class _____   Date _____

Drawing Atomic Models (cont.)

2. The atomic number of an atom equals the number of protons in the atom's nucleus. The sum of the number of protons and the number of neutrons in the nucleus is the mass number of an atom. In a data table like **Data Table 1** shown below, fill in the number of protons, the number of neutrons, the atomic number, and the mass number for the atom in **Figure 1** (Atom A).

3. Notice that **Figure 1** shows the first two electrons in Atom A in the first energy level, on the circle closest to the nucleus. The third electron is in the second energy level, on a circle farther from the nucleus. The first energy level of any atom can hold only up to two electrons. The second and third energy levels can each hold up to eight electrons. In a data table like **Data Table 2** shown below, fill in the total number of electrons, the number of energy levels, and the number of electrons in the highest energy level (the circle farthest from the nucleus) for Atom A.

4. On a separate piece of paper, draw a model of an atom that has 11 protons and 12 neutrons. Remember that the first energy level for electrons can hold only 2 atoms, while the second and third energy levels can hold up to 8. Label your atom "Atom B." The style of your model should be similar to the style of Atom A, shown in **Figure 1.**

5. Fill in your data table with the appropriate values for Atom B.

6. On your paper, draw a model of an atom that has an atomic number of 19 and a mass number of 39. Label this atom "Atom C." Fill in the appropriate values for this atom in your data table.

7. On your paper, draw a model of an atom that has an atomic number of 17 and a mass number of 35. Label this atom "Atom D." Fill in the appropriate values for this atom in your data table.

DATA TABLE 1

Atom	# of protons	# of neutrons	Atomic number	Mass number
Atom A	3	4	3	7
Atom B	11	12	11	23
Atom C	19	20	19	39
Atom D	17	18	17	35

Name _____ Class _____ Date _____

Drawing Atomic Models (cont.)

DATA TABLE 2

Atom	Total # of electrons	# of energy levels	# of electrons in highest energy level
Atom A	3	2	1
Atom B	11	3	1
Atom C	19	4	1
Atom D	17	3	7

PLACING ELEMENTS IN THE PERIODIC TABLE

8. Figure 2 shows part of a simple periodic table that is partially filled in. Copy this table onto a separate sheet of paper. Locate the proper places for Atoms A, B, C, and D on the table. On your copy of the table, write the name of the atom (A, B, C, or D) and the atomic number in the appropriate box for each atom.

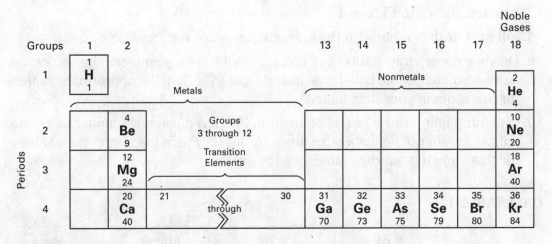

FIGURE 2

ANALYSIS

1. Recognizing patterns What value in the data tables do Atoms A, B, and C have in common? How is this similarity reflected in their positions in the periodic table?

Atoms A, B, and C have one electron in their highest energy level. All three

atoms are in the same column of the periodic table (Group 1).

Name _____ Class _____ Date _____

Drawing Atomic Models (cont.)

2. Recognizing patterns How does the period number of each atom compare with the number of energy levels in the atom?

The period number is the same as the number of energy levels in each atom.

_____

_____

_____

CONCLUSIONS

1. Drawing conclusions Examine a periodic table in your classroom or in your textbook. What are the proper names of Atoms A, B, C, and D?

Atom A = lithium; Atom B = sodium; Atom C = potassium;

Atom D = chlorine

_____

_____

_____

2. Drawing conclusions Which is more important in determining an element's chemical properties: its group or its period on the periodic table?

An element's group is more important than its period in determining its

chemical properties.

_____

_____

_____

3. Applying conclusions Find the element fluorine on a periodic table. How many protons does a fluorine atom have? How many electrons does a neutral fluorine atom have? Which one of the atoms that you studied in this experiment is most chemically similar to fluorine?

A neutral fluorine atom has nine protons and nine electrons. Atom D

(chlorine) is most chemically similar to fluorine.

_____

_____

_____

Name _____ Class _____ Date _____

Drawing Atomic Models (cont.)

EXTENSIONS

1. Research and communications Examine a periodic table. Notice that the second and third periods (the second and third horizontal rows) have a space between the second and third columns. This space allows the transition elements (whose atomic numbers are 21–30) to fit into the table in later periods. Research the number of electrons that can fit into energy levels higher than the third level. Write a short paragraph that describes how electrons fit into these higher levels and that explains why a space must be left in the lower periods on the periodic table.

Energy levels 4 and higher can hold up to 18 electrons. There is an additional orbital, called the d orbital, that can hold 10 electrons. For that reason, a space 10 columns wide must be left in the lower periods to allow the additional 10 elements to fit in the higher periods. (Note: An f orbital can hold 14 electrons in still larger atoms. Space for these elements in Periods 6 and 7 is usually accounted for by placing elements 58–71 and 90–103 below the rest of the periodic table.)

Teacher's Notes and Answers

Extracting Iron from Cereal

TIME REQUIRED
1 lab period

SKILLS ACQUIRED
Classifying
Experimenting
Inferring
Interpreting
Communicating

THE SCIENTIFIC METHOD

- **Make Observations:** In the procedure steps, students will observe the effects of magnets on several forms of iron.
- **Analyze the Results:** In Analysis questions 1–3, students will analyze the data and results from their experiments.
- **Draw Conclusions:** In Conclusions questions 1–4, students use their results to draw conclusions about which form of iron is found in cereal and in iron supplements.

Teacher's Notes
MATERIALS

You will need to make 20 magnetic stirring rods for this activity. For each stirring rod, place a magnet near one end of a pencil and attach the magnet with tape. Place each pencil-magnet assembly in the bottom of a sandwich bag, and seal the bag.

The cereal used in this lab must be iron fortified. In iron supplements, three capsules typically contain 18 mg of iron, which is the recommended daily allowance (RDA) set by the U.S. government.

ADDITIONAL MATERIALS

electrical tape, waterproof (1 roll)
magnets, 1 to 2 cm in diameter (20)
pencils, slightly shorter than the width of the bags (20)
sandwich bags, sealable and plastic (20)

SAFETY CAUTIONS

Students should not be allowed to eat the cereal or iron supplement capsules used in this lab.

TIPS AND TRICKS

This activity works best in groups of 2–3 students. You may begin the activity by showing students a bag of iron filings or a box of nails and telling them, "I found this in my cereal this morning!"

You may wish to discuss the difference between a mixture and a compound. This distinction is important when considering the absorption of iron by our bodies. Iron is most readily used by the body when it is in a compound, such as ferrous sulfate, ferrous fumarate, or ferrous gluconate. The iron in many dietary supplements is in compound form. Unfortunately, this form of iron also tends to oxidize when it is exposed to air, and oxidized iron is unusable. Iron that is mixed into certain fortified breakfast cereals is in a metallic state because metallic iron does not oxidize as easily. The metallic iron is converted to a ferrous compound by the hydrochloric acid in the stomach, where it is then absorbed by the body.

DISPOSAL INFORMATION

Students should dispose of any iron filings, cereal, and iron supplement capsules (and their contents) that are used in the lab. These materials may be disposed of in a trash can.

Name_____ Class_____ Date_____

Skills Practice Lab

Extracting Iron from Cereal

Introduction

You have probably walked down the cereal aisle in a supermarket and seen the phrase "Fortified with iron" on many of the boxes. That means iron has been added to the cereal. Why would iron be added to cereal? Iron helps carry oxygen to different parts of your body. Green, leafy vegetables are good sources of iron, but iron-fortified cereals are, too. How could you find out if there is iron in your cereal? One way you can tell is by using a magnet. Iron is found in two forms: as an element and as part of a compound. Elemental iron is attracted to magnets, while iron in a compound form is not. If the cereal contains elemental iron, the iron can be extracted with the magnet. In this experiment, you will attempt to extract iron from cereal and compare the iron extracted from cereal to iron found in iron supplement capsules.

OBJECTIVES

Compare the effects of a magnetic rod on metallic iron and on an iron compound.

Infer whether the iron in iron-fortified cereal is in a compound or in an elemental, metallic state.

MATERIALS

beakers, 500 mL (2)
graduated cylinder, 1 L
iron capsules, dietary (3)
iron-fortified cereal (90 g)
iron filings (small amount)

magnetic stirring rods, each in a small
plastic bag (2)
watch or clock
water

SAFETY

- Secure loose clothing and remove dangling jewelry. Don't wear open-toed shoes or sandals in the lab.

- Check the condition of glassware before and after using it. Inform your teacher of any broken, chipped, or cracked glassware because it should not be used.

- Do not pick up broken glass with your bare hands. Place broken glass in a specially designated disposal container.

Name _____ Class _____ Date _____

| *Extracting Iron from Cereal (cont.)*

Procedure

1. Move one of the stirring rods close to the iron filings. Observe what happens to the filings.

2. Remove any filings from the outside of the bag, and discard them.

3. Thoroughly crush the cereal. Place the cereal in the first beaker, and add 100 mL of water. Set the beaker aside. Do not eat the cereal.

4. Empty the contents of the iron capsules into the second beaker, and add 100 mL of water. Set the second beaker aside. Do not eat the iron supplement capsules.

5. Roll up the bags containing the stirring rods so that they are easier to hold. After the cereal has become soggy (10 to 15 minutes), use a stirring rod, magnet side down, to slowly stir the contents of each beaker for 5 minutes.

6. Carefully remove both stirring rods, and examine them closely.

Analysis

1. **Describing events** What happened to the iron filings when the stirring rod approached them?

   The iron filings moved or stuck to the rod because they were attracted to

   the magnet.

   _____

2. **Describing events** What did you see on the stirring rod from the first beaker?

   Little black particles were stuck to the stirring rod.

   _____

   _____

3. **Describing events** What did you see on the stirring rod from the second beaker?

   No particles were on the second stirring rod.

   _____

   _____

CONCLUSIONS

1. Interpreting information Based on your observations in step 1, are the iron filings made of elemental iron or of an iron compound? Explain your answer.

The iron in the filings must be elemental iron because they were attracted to

the magnet.

2. Drawing conclusions Based on your observations, do you think elemental iron is present in one or both of the beakers? Explain your answer.

Elemental iron is in the first beaker because the particles stick to the mag-

net. Also, the particles on the first stirring rod are dark and metallic, like

iron filings. No elemental iron is in the second beaker.

3. Drawing conclusions What form of iron was in the iron supplement capsules? Explain your answer.

The capsules must contain iron in the form of a compound because nothing

stuck to the magnet in the second beaker.

4. Applying conclusions If you dip a magnetic stirring rod into cereal and no iron sticks to the rod, does that mean there is no iron in the cereal? Explain your answer.

No; if iron does not stick to the magnet, it might mean that the iron is in the

form of a compound. Iron as a compound does not stick to magnets.

EXTENSIONS

1. Research and communications Find out what the terms *fortified* and *enriched* mean when they appear on nutritional food labels. Explain.

The term *fortified* on a nutritional label means that nutrients have been

added that were never in the food. One example is the addition of vitamin D

to milk. The term *enriched* means that nutrients have been added that were

once present in the food. Such nutrients are usually lost as a result of food

processing, such as making white bread from whole grains.

Teacher's Notes and Answers

Combining Elements

TIME REQUIRED
1 lab period

SKILLS ACQUIRED
Collecting data
Measuring
Organizing and analyzing data

THE SCIENTIFIC METHOD

- **Make observations:** In the Procedure and Analysis, students measure masses and observe changes in the appearance of substances in an evaporating dish as it is heated.

- **Analyze the results:** In the Analysis, students calculate a change in mass and determine whether the mass increases or decreases.

- **Draw conclusions:** In the Conclusions, students draw conclusions about why a change in mass occurred and how the chemical reaction took place.

- **Communicate results:** In the Analysis and Conclusions, students communicate results by providing written answers to the questions.

Teacher's Notes
MATERIALS

You may provide each lab group with a container of copper powder, or you may place a large container of copper powder near the balance.

SAFETY CAUTIONS

Students will be using open flame, glassware, and chemicals in this experiment. Review laboratory safety information thoroughly with students before they start the experiment.

DISPOSAL INFORMATION

Copper(II) oxide and any unused copper powder may be disposed of in a trash can. Evaporating dishes should be rinsed with water after the experiment is finished.

TECHNIQUES TO DEMONSTRATE

If you have not taught students how to use Bunsen burners or it has been a while since they used a burner, demonstrate proper lighting of a burner and explain the safety precautions that they should follow when working near an open flame.

Name _____ Class_____ Date_____

Skills Practice Lab

Combining Elements

Introduction

If you have ever left something metallic out in the rain, you may have returned to find the item spotted with rust. Rust forms when iron combines with oxygen in damp air to make the compound iron(III) oxide. The formation of rust from iron is an example of a synthesis reaction. A synthesis reaction is any reaction in which two or more substances combine to form a single compound. The resulting compound has different chemical and physical properties than the substances that compose the compound do. In this activity, you will synthesize, or create, copper(II) oxide from the elements copper and oxygen.

OBJECTIVES

Describe the change in appearance of copper when copper combines with oxygen to form copper(II) oxide.

Compute the change in mass that occurs in a synthesis reaction.

Explain why mass increases in a synthesis reaction.

MATERIALS

Bunsen burner or portable burner   ring stand and ring
copper powder   spark igniter
evaporating dish   tongs
metric balance   weighing paper
protective gloves   wire gauze

SAFETY

- Never taste, touch, or smell chemicals unless specifically directed to do so.

- Secure loose clothing and remove dangling jewelry. Don't wear open-toed shoes or sandals in the lab.

- Wear an apron or lab coat to protect your clothing when working with chemicals.

- Wear safety goggles when working around chemicals, acids, bases, flames, or heating devices. Contents under pressure may become projectiles and cause serious injury.

- Avoid wearing contact lenses in the lab.

- If any substance gets in your eyes, notify your instructor immediately and flush your eyes with running water for at least 15 minutes.

- Check the condition of glassware before and after using it. Inform your teacher of any broken, chipped, or cracked glassware because it should not be used.

Name _____ Class _____ Date _____

Combining Elements (cont.)

- In order to avoid burns, wear heat-resistant gloves when handling chemicals.
- If you are unsure of whether an object is hot, do not touch it.
- Avoid wearing hair spray or hair gel on lab days.

Procedure

1. Use the metric balance to measure the mass (to the nearest 0.1 g) of the empty evaporating dish. Record this mass in a table like the one shown below.

2. Place a piece of weighing paper on the metric balance, and measure approximately 10 g of copper powder. Record the mass (to the nearest 0.1 g) in the table. Caution: When working with copper powder, wear protective gloves.

3. Use the weighing paper to place the copper powder in the evaporating dish. Spread the powder over the bottom and up the sides as much as possible. Discard the weighing paper.

4. Set up the ring stand and ring. Place the wire gauze on top of the ring. Carefully place the evaporating dish on the wire gauze.

5. Place the Bunsen burner under the ring and wire gauze. Use the spark igniter to light the Bunsen burner. Caution: When working near an open flame, use extreme care.

6. Heat the evaporating dish for 10 min.

7. Turn off the burner, and allow the evaporating dish to cool for 10 min. Use tongs to remove the evaporating dish, and place it on the balance to determine the mass. Record the mass in your data table.

DATA TABLE

Object	Evaporating dish	Copper powder	Copper + evaporating dish after heating	Copper(II) oxide
Mass (g)				

ANALYSIS

1. Analyzing data Compute the mass of the product of the reaction—copper(II) oxide—by subtracting the mass of the evaporating dish from the mass of the evaporating dish and copper powder after heating. Record this mass in your data table.

Answers may vary.

Name _____ Class _____ Date _____

Combining Elements (cont.)

2. Examining data Did the mass of the substance in the dish increase or decrease as a result of heating the copper?

The mass of the substance in the dish increased.

_____

_____

_____

3. Describing events What evidence of a chemical reaction did you observe after the copper was heated?

The copper changed color, and the mass changed.

_____

_____

_____

CONCLUSIONS

1. Drawing conclusions Explain why a change in mass occurred as a result of the reaction.

A change in mass occurred because the copper combined with oxygen from

the air. Copper(II) oxide has more mass than copper does.

_____

2. Drawing conclusions How does the change in mass support the idea that this reaction is a synthesis reaction?

A synthesis reaction is a reaction in which two or more substances join to

form a new substance. The mass of the copper(II) oxide is greater than the

mass of the copper alone, so a synthesis reaction, in which the copper com-

bined with oxygen from the air, must have occurred, which resulted in an

increase in mass.

3. Interpreting information Where did the oxygen in this reaction come from?

The oxygen came from the air.

_____

_____

_____

Name _____ Class _____ Date _____

Combining Elements (cont.)

4. **Evaluating methods** Why was powdered copper rather than a small piece of copper used? (Hint: How does surface area affect the rate of the reaction?)

   Powdered copper has a larger surface area than a piece of copper does.

   Greater surface area increases the rate of the reaction because more copper

   is exposed to oxygen.

5. **Evaluating methods** Why was the copper heated?

   The copper was heated because the formation of copper(II) oxide requires

   an input of energy to get started. Heating also speeds the reaction once the

   reaction is underway.

6. **Applying conclusions** Sometimes, the copper bottoms of cooking pots turn black after the pots are used. How is that similar to the results you obtained in this lab?

   The copper(II) oxide synthesized in this experiment is the same black pow-

   der that appears on copper pots.

   _____

EXTENSIONS

1. **Research and communications** Research methods for preventing the formation of rust on automobiles. Explain why it is necessary to paint cars and to patch nicks in the paint. Also, describe at least one additional method for preventing rust on automobiles.

   Painting a car helps prevent rust from forming by creating a barrier between

   the iron of the car and the oxygen. If the iron and oxygen are not in contact,

   they cannot react to form rust.

Teacher's Notes and Answers

Separating Substances in a Mixture

TIME REQUIRED
1–2 lab periods

SKILLS ACQUIRED
Classifying
Collecting data
Communicating
Designing experiments
Experimenting
Measuring
Organizing and analyzing data
Predicting

THE SCIENTIFIC METHOD

- **Make Observations:** In steps 1–3 of the Procedure, students observe the physical properties of various substances. In steps 6–8 of the Procedure, students observe how their methods to separate the substances from a mixture do or do not work.

- **Form a Hypothesis:** In step 4 of the Procedure, students propose methods for separating substances from a mixture.

- **Analyze the Results:** In the Analysis, students state the results of the experiment.

- **Draw Conclusions:** In the Conclusions, students draw conclusions about the effectiveness of their methods and suggest possible improvements.

- **Communicate the Results:** In the Analysis and Conclusions, students communicate results by providing written answers to the questions.

Teacher's Notes
MATERIALS

Prepare a bowl of heterogeneous mixture for each group of students. Each bowl should contain 100 g of cracked (not ground) black pepper, 500 g (500 mL) of water, 200 g of sand, 500 g of sugar, 50 g of iron filings, and 150 g of small nuts (such as peanuts). Thoroughly mix the contents of each bowl.

Also prepare small dishes that each contain a different substance. You may provide these small dishes to each lab group or have larger dishes in a central location for all the students to share. The students will use these isolated substances to observe the physical properties of the substances before developing their own methods for extracting the substances from the mixture.

SAFETY CAUTIONS

Read carefully through the methods proposed by students in step 3 of the Procedure. Be very sure that the methods are safe before allowing students to proceed.

Review general lab safety with your students before beginning this experiment.

Whenever possible, provide students with an electric hot plate instead of an open flame. Remind students to wear oven mitts, lab aprons, and goggles while using hot plates. Do not allow students to place any of the substances in this experiment directly on a hot plate. Make sure all glassware to be used on hot plates is resistant to heat.

DISPOSAL INFORMATION

Solid substances extracted from the mixture can be disposed of in a trash can. Water can be poured out in a sink.

TIPS AND TRICKS

This activity works best when students work in pairs. You may wish to have students review some of the physical properties of matter before they begin this lab. This review may help students to come up with a procedure for separating the substances.

Read carefully through the methods students propose in step 3 of the Procedure. Be very sure the methods are safe before allowing the students to proceed. If they have trouble coming up with methods or if their methods look like they probably will not work, you can prompt the students with hints or suggestions about a method or part of a method for extracting one of the substances. Point out to students that the order in which they do the extractions is important. The substances in the chart are presented in a recommended order.

Encourage students to measure the volume of the water while it contains the dissolved sugar and the iron filings. Students may be concerned that the sugar and iron filings will increase the water's volume. Assure students that adding the sugar and iron filings changes the volume only slightly, and direct the students to record their concerns in response to question 2 of the Conclusions.

Students may need help in developing a technique to separate the iron filings from the sugar. You might suggest that students pour the water, sugar, and iron filings into a beaker and then heat the beaker until the water evaporates. Warn students not to burn or caramelize the sugar.

Name _____ Class _____ Date _____

Skills Practice Lab

Separating Substances in a Mixture

Introduction

A mixture is a combination of one or more pure substances. A heterogeneous mixture is a mixture in which the substances are not uniformly mixed. Because the substances in a mixture may have different physical properties, these properties can be used to separate the substances from the mixture. In this experiment, you will examine several substances to determine their physical properties. Then you will develop your own methods for extracting these substances from a mixture, and carry out an experiment to test those methods.

OBJECTIVES

Identify key physical properties that could be used to extract a substance from a mixture.

Develop methods for extracting substances from a mixture.

Evaluate the effectiveness of your methods, and recommend improvements.

MATERIALS

balance	nuts, small
beaker, 1 L	oven mitts, (1 pair)
bowl, empty	pepper, cracked
bowl with mixture of substances	plastic-foam cups (6)
craft stick	sand
filter screens (2)	sieve or colander, small
graduated cylinder	spoon, plastic
hot plate	sugar
iron filings	towel, small
magnet, strong	water

SAFETY

- Always use caution when working with chemicals.

- Never taste, touch, or smell chemicals unless specifically directed to do so.

- Secure loose clothing and remove dangling jewelry. Don't wear open-toed shoes or sandals in the lab.

- Wear an apron or lab coat to protect your clothing when working with chemicals.

- If a spill gets on your skin or on your clothing or in your eyes, rinse it immediately, and alert your instructor.

Name _____ Class _____ Date _____

Separating Substances in a Mixture (cont.)

- Wear safety goggles when working around chemicals, acids, bases, flames, or heating devices. Contents under pressure can become projectiles and cause serious injury.

- Avoid wearing contact lenses in the lab.

- If any substance gets in your eyes, notify your instructor immediately and flush your eyes with running water for at least 15 minutes.

- Check the condition of glassware before and after using it. Inform your teacher of any broken, chipped, or cracked glassware because it should not be used.

- Do not pick up broken glass with your bare hands. Place broken glass in a specially designated disposal container.

- In order to avoid burns, wear heat-resistant gloves when handling chemicals.

- If you are unsure whether an object is hot, do not touch it.

- Avoid wearing hair spray or hair gel on lab days.

- Dispose of all sharps (broken glass and other contaminated sharp objects) and other contaminated materials (biological and chemical) in special containers as directed by your instructor.

Procedure

1. Learning the physical properties of a substance can help you decide how to separate it from other substances. Observe the physical properties of each of the following substances: cracked pepper, iron filings, nuts, sand, sugar, and water. Caution: Do not taste or eat any of these substances.

2. For each substance, answer the following questions in the Physical properties column of the chart on the next page: Does this substance dissolve in water? Does it float? Is it magnetic? Are its pieces large or small relative to the pieces of other substances? Is it a solid, a liquid, or a gas?

3. Look over the properties you recorded in step 2. For each substance, determine which characteristic would help you best distinguish it from the other substances. Choose from among the following properties: size, shape, density, state of matter, solubility, and magnetic attraction. Record a distinguishing characteristic on the chart for each substance.

Name _____ Class _____ Date _____

Separating Substances in a Mixture (cont.)

DATA TABLE Separation of Substances

Substance	Make Observations		Form Hypotheses		Conduct an Experiment	
	Physical properties	Distinguishing characteristic	Method of separation		Check when done	Mass extracted (g)
Pepper	small, but larger than sand; does not dissolve in water; floats; not magnetic	density	Allow the mixture to settle. Scrape the layer of floating pepper from the water with a craft stick.		✓	100
Nuts	largest substance; do not dissolve in water; do not float; not magnetic	size	Lift the nuts with the spoon. Dry the nuts on towels.		✓	150
Sand	small; does not dissolve in water; does not float; gritty texture; not magnetic	shape and size	Place the sieve over the bowl. Put a piece of filter screen in the sieve. Pour the mixture into the filter. Catch and save the water. Dry the solids.		✓	200
Iron filings	small; magnetic; do not dissolve in water; do not float	magnetic	Pass a magnet through the dry mixture.		✓	50
Water	liquid	liquid state	Measure mass of water remaining after removing all other substances.		✓	500
Sugar	small; dissolves in water; not magnetic	solubility in water	Pour water into a beaker. Boil the water. When the water is gone, cool the remaining substances.		✓	500

Name _____ Class _____ Date _____

Separating Substances in a Mixture (cont.)

4. You have been given a bowl containing a mixture of all the substances you have observed: pepper, iron filings, nuts, sand, sugar, and water. Formulate hypotheses about how you could separate each substance from the rest of the mixture using the materials provided for you in the lab. It may help to look over the materials list. If you are working in a group, discuss your proposed methods with the others in the group and decide together on a method for each substance. Describe your proposed methods in the "Method of separation" column in the table. The first row in the chart is filled in to help you get started.

5. Have your teacher approve the methods you have described, then conduct an experiment to test your methods.

6. Measure the mass of an empty cup, and record the mass.

7. Follow your plans to separate each substance from the mixture. If your plans change, be sure to note the changes in the method column on the chart. Be careful not to spill any of the substances on the floor as you work. Protect your hands with oven mitts when working with the hot plate, and be careful not to burn any of the substances. Never place any of the substances directly on the hot plate. As you extract the substances from the mixture, store each substance in a different cup, and label each cup.

8. When you have finished extracting all the substances from the mixture, measure the mass of each cup and its contents. For each measured mass, subtract the mass of the empty cup that you measured in step 6, and record your results in the last column on the chart.

ANALYSIS

1. Organizing data List the substances in the mixture in order from the least mass extracted to the greatest mass extracted.

iron filings, pepper, nuts, sand, water, and sugar _____

_____

_____

2. Classifying Was this mixture a homogeneous mixture or a heterogeneous mixture? Was it a suspension, a colloid, an emulsion, or a solution?

The mixture was heterogeneous and was a suspension. _____

_____

_____

_____

Name _____ Class _____ Date _____

Separating Substances in a Mixture (cont.)

CONCLUSIONS

1. **Evaluating results** Were your measurements of the masses of the substances in the mixture accurate? Why or why not?

   Not all mass measurements were completely accurate. Possible reasons may

   include the following: some of the pepper stuck to the craft stick and the

   nuts; the nuts were still damp and were coated with sand; the sand was still

   damp; and some iron filings were buried in the sand.

2. **Evaluating methods** How could you improve your methods to better separate each substance?

   Accept all reasonable answers. Possible improvements include the following:

   removing all of the pepper from the craft stick and the nuts; waiting until

   the nuts are completely dry and then brushing the rest of the sand off the

   nuts; evaporating the water from the sand; removing the sand from the nuts

   and the filter paper; using a stronger magnet to pull all of the filings out of

   the sand; squeezing the filter to remove some of the water that the filter

   absorbed; and letting the water evaporate more slowly.

EXTENSIONS

1. **Research and communications** A centrifuge is a device that uses centrifugal force to separate substances of different masses. Centrifuges are often used by biologists and biochemists to separate parts of cells or biological molecules of different masses. Research how centrifuges work, and write a short paragraph explaining what you learn.

   A centrifuge contains a chamber or chambers for holding a solution. When
   the centrifuge is turned on, it rotates at high speed. Heavier particles in the
   solution are drawn by centrifugal force to the end of the test tube that is
   farther from the center of the centrifuge. Lighter particles form a layer
   closer to the center or remain dissolved in the solution until the speed of the
   centrifuge is increased.

Teacher's Notes and Answers

Modeling Radioactive Decay with Pennies

TIME REQUIRED
1 lab period

SKILLS ACQUIRED
Classifying
Collecting data
Recognizing patterns
Interpreting
Organizing and analyzing data

THE SCIENTIFIC METHOD

- **Make Observations:** In the Procedure, students observe the number of pennies that come up as heads or tails when shaken and poured out of a container.

- **Analyze the Results:** In the Analysis, students answer questions about the data and plot a graph of the number of remaining coins versus the number of shakes.

- **Draw Conclusions:** In the Conclusions, students use the pennies as a model, determine the half-life of the pennies, and compare the decay of the pennies to the radioactive decay of carbon-14.

- **Communicate Results:** In the Analysis and Conclusions, students communicate results by providing written answers to the questions.

Teacher's Notes

MATERIALS

An empty coffee can or a one-quart plastic storage canister will work well as the container. You may wish to provide each group with a large, shallow pan or box for pouring the pennies into. This pan or box will minimize the risk of pennies spilling or rolling onto the floor. You may arrange to get large numbers of pennies from a local bank.

SAFETY CAUTIONS

A container of 100 pennies is fairly heavy and can present a hazard if dropped. Remind students not to wear open-toed shoes or sandals in the lab. Pennies spilled on the flooer can present a slipping hazard. Instruct students to pick up any spilled coins immediately.

DISPOSAL INFORMATION

You may exchange the pennies for other currency at a local bank, or you may keep the pennies for use in future experiments.

Name _____ Class_____ Date_____

Skills Practice Lab

Modeling Radioactive Decay with Pennies

Introduction

Imagine existing more than 5000 years and still having more than 5000 to go! That is exactly what the unstable element carbon-14 does. Carbon-14 is an unstable isotope of carbon. Carbon-14 is used in the radioactive dating of material that was once alive, such as fossil bones. Every 5730 years, half of the carbon-14 in a fossil specimen decays or breaks down into the stable element nitrogen-14. In the following experiment you will see how pennies can be a model for the same kind of decay.

OBJECTIVES

Discover how the number of coins remaining after shaking, pouring, and selecting for tails changes with each shake.

Graph the number of coins remaining as a function of the number of shakes.

Compare the graph of the number of coins remaining to a graph of radioactive decay.

MATERIALS

containers with covers, large (2)   pennies (100)

SAFETY

• Secure loose clothing and remove dangling jewelry. Don't wear open-toed shoes or sandals in the lab.

Procedure

1. Place 100 pennies in a large, covered container. Shake the container several times and remove the cover. Carefully empty the container on a flat surface, and make sure the pennies don't roll away.

2. Remove all the coins that have the head side of the coin turned upward, and place them in a separate container. In a data table like the one on the next page, record the number of pennies removed and the number of pennies remaining.

Name _____ Class _____ Date _____

Modeling Radioactive Decay with Pennies (cont.)

3. Place the remaining pennies (with the tail side showing) back into the original container. Shake the container, and empty it onto the flat surface again. Sort out the pennies, and record data as in step 2. Remember to remove only the coins showing heads. Repeat this process until no pennies are left in the container. If the process requires more than nine steps, extend your data table as needed.

4. When you are finished, return all pennies to the original container, and clean up your work area.

DATA TABLE

Shake number	Number of coins remaining	Number of coins removed
1		
2		
3		
4		
5		
6		
7		
8		
9		

ANALYSIS

1. **Constructing graphs** Draw a graph to plot your data. Label the x-axis "shake number," and label the y-axis "Pennies remaining." Using data from your data table, plot the number of coins remaining after each shake.

   Graphs should be clearly labeled and should show a roughly exponential

   decrease in the number of pennies such that the number is reduced by about

   half with each shake.

2. **Examining data** How many shakes did it take before the number of pennies remaining was about one half the original number of pennies (about 50)? How many shakes did it take before the number of pennies remaining was about one-fourth the original number of pennies (about 25)?

   The number of pennies was about 50 after one shake. The number was about

   25 after two shakes.

   _____

Name _____ Class _____ Date _____

Modeling Radioactive Decay with Pennies (cont.)

CONCLUSIONS

1. Drawing conclusions What is the half-life of the pennies in this experiment, in numbers of shakes?

The half-life of the pennies in this experiment is one shake.

_____

_____

2. Analyzing graphs Examine the graph below. Compare the graph you have made for pennies with the graph for carbon-14. Explain any similarities that you see between the graphs.

Half-Life of Carbon-14

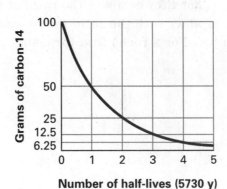

Number of half-lives (5730 y)

The graphs should be very similar in shape. With each half-life and each

shake, the number of pennies remaining is reduced by approximately half.

_____

3. Analyzing conclusions Imagine that you have found a fossilized leg bone of some unknown mammal. Based on the size of the bone, you determine that it should have contained about 100 g of carbon-14 when the animal was alive. The bone now contains about 12.5 g of carbon-14. How old is the bone?

The bone is about 17000 years old.

_____

_____

_____

Name _____ Class _____ Date _____

Modeling Radioactive Decay with Pennies (cont.)

EXTENSIONS

1. **Research and communications** Carbon-14 is used to date materials as old as about 60000 years. However, the age of Earth is thought to be 4.5 billion years, and life is thought to have existed on Earth for close to 4 billion years. These dates have been determined in part using radioactive dating methods. Research radioactive dating methods that can measure ages much older than 60000 years. Write a paragraph explaining at least two such methods. Include the names of the radioactive elements, the names of the elements they decay into, and the half-life of the reactions.

Examples of other radioactive dating methods include potassium-argon dating (^{40}K − ^{40}Ar with a half-life of 1.2 billion years) and uranium-lead dating (^{235}U − ^{206}Pb with a half-life of 700 million years). All of these methods are similar in that they compare the ratio of the radioactive material to its decay product with the ratio that would have existed when the object (such as a fossil or a rock) first formed.

Teacher's Notes and Answers

Testing Reaction Time

TIME REQUIRED
1 lab period

SKILLS ACQUIRED
Collecting data
Communicating
Experimenting
Recognizing patterns
Measuring
Organizing and analyzing data

THE SCIENTIFIC METHOD

- **Make Observations:** In the Procedure, students measure falling distance and falling time.

- **Analyze the Results:** In the Analysis, students calculate average falling distance, average falling time, and reaction time.

- **Draw Conclusions:** In the Conclusions, students compare results derived using different methods and evaluate the methods in the experiment.

- **Communicate Results:** In the Conclusions, students communicate results by providing written answers to the questions.

Teacher's Notes
MATERIALS

This experiment works best in groups of two. If you do not have enough metersticks for groups of two, you can use larger groups and each person can take a turn testing reaction time.

SAFETY CAUTIONS

Make sure students perform this experiment in a clear area so that the metersticks will not knock anything over. Make sure students are not wearing sandals or open-toed shoes. There is a chance that a meterstick could fall on their toes.

Name _____ Class _____ Date _____

Skills Practice Lab

Testing Reaction Time

Introduction

Objects in free fall near Earth's surface accelerate under the influence of gravity. This acceleration is constant, but it may be reduced somewhat by air resistance. Near Earth's surface, the acceleration due to gravity is 9.8 m/s^2. You can use this acceleration and a measurement of falling distance to calculate the time that an object is in free fall.

Predicting human reaction time is more complicated than predicting a simple physical event such as the time an object takes to fall. Human reaction time depends on many factors, such as the time a signal takes to go from your eyes to your brain, the time your brain takes to process the signal, and the time your brain takes to signal a reaction. Although the steps in reaction time are complicated, you can determine overall reaction time by using simple measurements.

In this lab, you will test your reaction time by measuring how far a meterstick falls before you can catch it. You will measure the distance that the meterstick falls and will then use an equation based on free-fall acceleration to determine the amount of time the meterstick took to fall. You will also measure the amount of time the meterstick falls by using a stopwatch and will compare the measured reaction time with the calculated reaction time.

OBJECTIVES

Calculate reaction time of a falling object by using falling-distance data.

Compare reaction times measured by using a stopwatch with reaction times calculated from falling-distance data.

Evaluate the methods used to determine reaction time in the experiment.

MATERIALS

calculator   paper
meterstick   stopwatch

SAFETY

• Secure loose clothing and remove dangling jewelry. Don't wear open-toed shoes or sandals in the lab.

Name _____ Class _____ Date _____

Testing Reaction Time (cont.)

Procedure

1. Have each person in the group prepare a data table like the one on the next page. Each person should write his or her name at the top of the table.

2. Work in pairs. Hold the meterstick vertically with the zero end down. Have your partner stand in front of you with his or her thumb and forefinger open about an inch and a half apart and lined up with the bottom (zero end) of the meterstick.

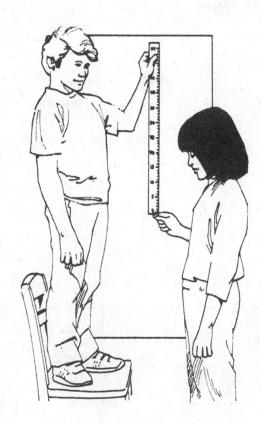

3. Drop the meterstick. Your partner should try to catch the meterstick between his or her thumb and forefinger as quickly as possible. Once your partner catches the meterstick, you can measure how far the meterstick fell by reading the point on the meterstick that your partner grabbed. Convert this distance to meters, and record it in your data table.

4. Repeat steps 2–3 nine more times, for a total of 10 trials. Record all falling distances in your data table.

5. Repeat steps 2–3 again, but this time use a stopwatch. Try to start the stopwatch at the same time that you release the meterstick, and stop the stopwatch as soon as your partner catches the meterstick. Repeat this process until you have completed 5 trials. Record all times in the lower part of your data table. You do not need to measure or record distance for these time trials.

6. Change places with your partner, and repeat steps 2–5. Record all falling distances in another data table. If you are working in a group of more than two people, continue repeating steps 2–5 until everyone has had a chance to test reaction time.

Name _____ Class _____ Date _____

| *Testing Reaction Time (cont.)*

DATA TABLE

Name: _____			
Distance data		**Time data**	
Trial	**Distance (m)**	**Trial**	**Time (s)**
1		11	
2		12	
3		13	
4		14	
5		15	
6		Average time (s)	
7			
8			
9			
10			
Average distance (m)			
Reaction time (s)			

ANALYSIS

1. **Organizing data** For each person in the group, calculate the average falling distance by adding the distance for the first 10 trials and then dividing by 10. Record the average in your data table.

   Answers may vary depending on the data. _____

   _____

   _____

2. **Organizing data** For each person in the group, calculate reaction time from the average distance by using the following formula:

$$reaction\ time = \sqrt{\frac{2 \times average\ distance}{9.8\ \text{m/s}^2}}$$

   Answers may vary depending on the data. _____

   _____

   _____

Name _____ Class _____ Date _____

Testing Reaction Time (cont.)

3. **Organizing data** For each person in the group, calculate the average falling time by adding the time for the last 5 trials and then dividing by 5. Record the average in your data table.

Answers may vary depending on the data.

_____

_____

CONCLUSIONS

1. **Evaluating methods** How does the average falling time you calculated by using data from the stopwatch compare with the reaction time you calculated by using the average falling distance? Should these values be the same? Why might these values be different?

In most cases, the average falling time calculated by using the stopwatch is

longer than the reaction time calculated by using distance data. In theory,

these values should be the same. However, using the stopwatch introduces

additional variables, such as the reaction time of the person who is using the

stopwatch.

2. **Evaluating results** How does your reaction time compare with those of other members of your group? Who has the fastest reaction time?

Answers may vary.

_____

_____

_____

3. **Evaluating data** How did your reaction time change over the first 10 trials? Do you see any evidence of improvement with practice?

Answers may vary. Some students may note an improvement over the trials,

while other students may not.

_____

_____

Name _____ Class _____ Date _____

Testing Reaction Time (cont.)

4. Evaluating methods Why did each of you do 10 trials to find an average distance instead of doing just 1 trial? Do you think your results would be more accurate if you did even more trials? Explain your answer.

Taking an average of 10 trials tends to even out the differences in individual

trials. Doing more trials should make the results more accurate (but if you

do too many trials, human error starts to increase because of fatigue).

EXTENSIONS

1. Designing experiments Repeat this experiment, but blindfold the person who is catching the meterstick. As soon as you drop the meterstick, signal your blindfolded partner by using your voice. Repeat this process for several trials. Then, repeat the process for several more trials, but signal your blindfolded partner by tapping your partner on the arm or shoulder. You do not need to use the stopwatch; use only falling-distance data. Which of these signaling methods produces the fastest reaction time? How do these reaction times compare with the reaction time when the catcher is not blindfolded?

This exercise can be a fun way for students to study reaction time in greater

depth. Make sure that students are not wearing sandals or open-toed shoes

(the meterstick could land on their feet). In most cases, students will find

that tapping the catcher leads to a faster reaction time than using the voice

does. The brain reacts faster to touch than to voice cues.

Teacher's Notes and Answers

Exploring Work and Energy

TIME REQUIRED
1 or 2 labs

SKILLS ACQUIRED
Collecting data
Communicating
Experimenting
Interpreting
Measuring
Organizing and Analyzing data

THE SCIENTIFIC METHOD

• **Make Observations:** In Procedure steps 3–6, 9–12, and 16–19, students will observe forces measured on a force meter.

• **Analyze the Results:** In Analysis questions 1–8, students will analyze the data and results from their experiments.

• **Draw Conclusions:** In Conclusions questions 1–8, students will use their results to draw conclusions about the work required to move mass by using different methods.

Teacher's Notes
MATERIALS

The hooked mass sets should include 1 kg and 0.2 kg masses.

Each lab group should have two different spring scales. One should be 20 N _0.5 N and the other should be 2.5 N _0.1 _. Students should use the smaller scale unless the force required to move the mass is greater than 2.5 N.

SAFETY CAUTIONS

Students should secure all masses and apparatus. Students should perform this experiment in a clear area, out of the way of traffic and emergency exits. Students should handle only one mass at a time to decrease the likelihood of dropping the mass.

TIPS AND TRICKS

Students can perform this lab by themselves or in groups of two or more.

Because each section of this exercise builds on the previous section, it is more efficient to leave the apparatus in place throughout the lab unless directed to remove it.

• **Step 11:** A supervised student may step on a sturdy chair to raise the mass more easily to this height.

Exploring Work and Energy (cont.)

- **Step 13:** Ideally, the base of the inclined plane should rest against a wall to ensure that the inclined plane does not fall.

- **Step 14:** Make sure that students measure both the vertical distance and the distance along the inclined plane. Students should measure vertical distances to the top of the inclined plane. Remind students to use the distance along the inclined plane in their calculations of work done on the mass.

TECHNIQUES TO DEMONSTRATE

Set up an inclined-plane apparatus for the students to use as an example.

Name _____ Class _____ Date _____

Skills Practice Lab

Exploring Work and Energy

Introduction

Common sense may tell you that pushing or pulling a heavy object up a ramp is easier than lifting the object straight up. Pushing or pulling an object up a ramp usually does require less force than lifting it does, but these actions do not always require less work. Work in the scientific sense is calculated as the force applied to an object times the distance the object moves. When you push or pull an object up a ramp, you increase the total distance that you have to move the object in order to raise it to a given height. Furthermore, you have to overcome the force of friction between the object and the ramp.

   In this experiment, you will measure the forces required to move objects across a level surface, to lift the objects straight up, and to pull the objects up an inclined plane. You will also calculate the work done on the objects in these three cases. Then, you will compare the amount of force and the amount of work required in each case.

OBJECTIVES

Measure the force required to move a mass over a certain distance.

Compute the work done on a mass.

Compare the work done on a mass and the force required to move the mass by using different methods.

MATERIALS

clamps	inclined plane
cord, 1.00 m	masking tape
force meters, spring scales (2)	meterstick
hooked masses, one set	stopwatch

Safety

• Secure loose clothing and remove dangling jewelry. Don't wear open-toed shoes or sandals in the lab.

Name _____ Class _____ Date _____

Exploring Work and Energy (cont.)

Procedure
PULLING MASSES

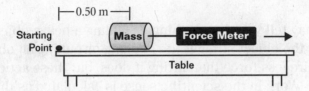

1. At one edge of the tabletop, place a tape mark to represent a starting point. From this mark, measure exactly 0.25 m and 0.50 m. Place a tape mark at each measured distance.

2. Securely attach the 1 kg mass to one end of the cord and the force meter to the other end. The force meter will measure the force required to move the mass through different displacements.

   Set up the apparatus, and attach all masses securely. Perform this experiment in a clear area. Swinging or dropped masses can cause serious injury.

3. Place the mass on the table at the starting point. Hold the force meter parallel to the tabletop so that the cord is taut between the force meter and the mass. Carefully pull the mass at a slow, constant speed along the surface of the table to the 0.25 m mark (this may require some practice). As you pull, observe the force measured on the force meter.

4. Record the force and distance in a table using the appropriate SI units (newtons and meters).

5. Repeat steps 3 and 4 for a distance of 0.50 m.

6. Repeat steps 3–5 with a 0.2 kg mass.

LIFTING MASSES

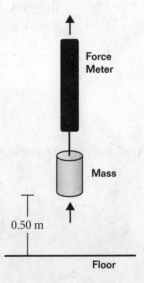

7. Using masking tape, secure a meterstick vertically against the wall so that the 0.00 m end touches the floor.

Name _____ Class _____ Date _____

Exploring Work and Energy (cont.)

8. Securely attach the 1 kg mass to one end of the cord and the force meter to the other end.

9. Place the mass on the floor beside the meterstick. Hold the force meter parallel to the wall so that the cord is taut between the force meter and the mass. Carefully lift the mass vertically at a slow, constant speed to the 0.25 m mark on the meterstick. Be sure that the mass does not touch the wall during any part of the process. As you lift, observe the force measured on the force meter. Be careful not to drop the mass.

10. Record the force and distance in your notebook using the appropriate SI units.

11. Repeat steps 9 and 10 for a vertical distance of 0.50 m.

12. Replace the 1 kg mass with the 0.2 kg mass, and repeat steps 9–11.

DISPLACING MASSES ON AN INCLINED PLANE

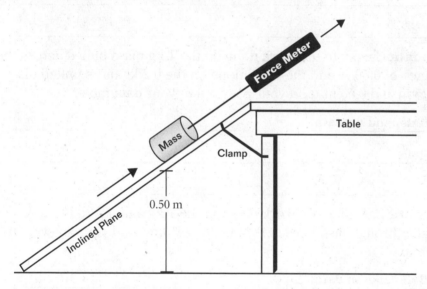

13. Carefully clamp an inclined plane to the tabletop so that the base of the inclined plane rests on the floor. Make sure the inclined plane is in a location where it will not obstruct traffic or block aisles or exits.

14. Measure vertical distances of 0.25 m and 0.50 m above the level of the floor. Use masking tape to mark each level on the inclined plane. Also measure the distance along the inclined plane to each mark. Record all distances in your notebook using the appropriate SI units. Be sure to label the vertical distance and the distance along the inclined plane.

15. Attach the 1 kg mass to the lower end of the cord and the force meter to the other end.

16. Place the mass at the base of the inclined plane. Hold the force meter parallel to the inclined plane so that the cord is taut between the force meter and the mass. Carefully pull the force meter at a slow, constant speed parallel to the surface of the inclined plane until the mass has reached the vertical 0.25 m

Name _____ Class _____ Date _____

Exploring Work and Energy (cont.)

mark on the inclined plane. As you pull, observe the force measured on the force meter.

17. Using the appropriate SI units, record the force and distance in your notebook.

18. Repeat steps 16 and 17 for a vertical distance of 0.50 m.

19. Repeat steps 16–18 for the 0.2 kg mass.

ANALYSIS

1. Organizing data Calculate the work done on the 1 kg mass pulled across the table a distance of 0.25 m. Use the work equation, $W = F \times d$. Add your answer to your data table.

Answers will depend on data.
_____

_____

_____

2. Organizing data Calculate the work done on the 1 kg mass pulled across the table a distance of 0.50 m and the work done on the 0.2 kg mass pulled distances of 0.25 m and 0.50 m. Add your answers to your data table.

Answers will depend on data.
_____

_____

_____

3. Organizing data Calculate the work done on the 1 kg and 0.2 kg masses lifted vertically through distances of 0.25 m and 0.50 m. Add your answers to your data table.

Answers will depend on data.
_____

_____

_____

4. Organizing data Calculate the work done on the 1 kg and 0.2 kg masses pulled up the inclined plane through vertical distances of 0.25 m and 0.50 m. In each calculation, remember to use the distance that the mass was pulled along the inclined plane, not the vertical distance. Add your answers to your data table.

Answers will depend on data.
_____

_____

Name _____ Class _____ Date _____

Exploring Work and Energy (cont.)

5. Analyzing data In each of the three setups, did you exert the same force on the 1 kg mass as you did on the 0.2 kg mass to move them an equal distance?

No; the 1 kg mass required more force.

6. Analyzing data In each of the three setups, did moving the mass 0.50 m require more force than moving the same mass 0.25 m did?

No; the same amount of force was constantly applied.

7. Analyzing results In each of the three setups, did you do more work on the 1 kg mass or on the 0.2 kg mass to move the masses a distance of 0.50 m?

More work was done on the 1.0 kg mass.

8. Analyzing results In each of the three setups, did you do more work on the 1 kg mass when you moved it a distance of 0.25 m or when you moved it a distance of 0.5 m?

More work was done on the mass when moving it 0.50 m.

CONCLUSIONS

1. Interpreting information What force did you pull against when you pulled the masses horizontally across the table?

the force of friction

Name _____ Class _____ Date _____

Exploring Work and Energy (cont.)

2. Interpreting information What force did you pull against when you lifted the masses vertically?

the force of gravity

_____

_____

3. Interpreting information What forces did you pull against when you pulled the masses up the inclined plane?

the forces of friction and gravity

_____

4. Drawing conclusions Compare your data on pulling the 1 kg mass across the table to your data on lifting the 1 kg mass vertically. Which force was greater, the force of friction between the mass and the table or the force of gravity acting on the mass?

The force of gravity was greater. (In some cases, the force of friction may be

greater.)

_____

5. Drawing conclusions Compare your data on lifting the 1 kg mass vertically through 0.50 m to your data on pulling the 1 kg mass up the inclined plane to a height of 0.50 m. In which case did you exert a greater force, and in which case did you do more work on the mass?

More force was exerted to lift the mass vertically. More work was done when

the mass was pulled up the inclined plane.

_____

6. Defending conclusions You should have found that you did more work on the mass to pull it up the inclined plane to a vertical height of 0.50 m than you did to lift it vertically to the same height (if you did not get that result, check your calculations or ask your teacher to look over your data). Explain why this is the case.

Although the vertical height was the same in both cases, the mass was moved

through a greater total distance when it was pulled up the inclined plane. In

addition, pulling the mass up the inclined plane required pulling against the

force of friction in addition to pulling against the force of gravity.

Name _____ Class _____ Date _____

Exploring Work and Energy (cont.)

7. Making predictions How could you adjust the inclined plane so that moving the mass through the same vertical displacement would require less force?

Use a longer ramp, and decrease the angle between the ramp and the floor.

_____

_____

8. Applying conclusions Would decreasing the angle of the ramp to the floor increase or decrease the amount of work done to move the mass to a vertical height of 0.50 m? Explain your answer.

Decreasing the angle would increase the amount of work because the force

of friction would be increased.

_____

EXTENSIONS

1. Designing experiments Design an experiment to test the amount of work done and the amount of force required to pull a mass up an inclined plane at different angles to the ground. Make a hypothesis about how changing the angle will affect the work done and the force required to lift the mass to a given vertical height.

A good experimental design would include pulling a given mass to a given vertical height by using an inclined plane at several different angles. As the angle decreases, the force required would decrease, but the total work would increase.

Teacher's Notes and Answers

Energy Transfer and Specific Heat

TIME REQUIRED
1 lab period

SKILLS ACQUIRED
Collecting data
Communicating
Experimenting
Measuring
Organizing and Analyzing data
Predicting

SCIENTIFIC METHOD

- **Form a Hypothesis:** In step 1 of the procedure, students will form a hypothesis about how they expect the temperatures to change.

- **Analyze the Results:** In Analysis questions 1–3, students will analyze the data and results from their experiments.

- **Draw Conclusions:** In Conclusions questions 1–6, students use their results to draw conclusions about energy transfer and specific heat.

Teacher's Notes
MATERIALS

Nails should be made of iron and should be short enough to be covered by water in the plastic-foam cups without causing the cups to be too full.

It is not necessary to heat the water. Hot tap water will make the experiment both faster and safer. If you want to use hotter water, heat the water for all stations on a single hot plate and emphasize that the students should wear gloves when handling water and nails.

If you want students to perform the experiments they design in the Extensions section, you can provide them with nails made of a different metal or of a metallic alloy.

SAFETY CAUTIONS

Caution students about working with hot water and nails. Instruct students to wear gloves when handling water and nails.

Remind students about safety and cleanup issues when using thermometers.

If you are concerned about having nails in the lab, you can clip and file the tips of the nails so that they are not sharp.

DISPOSAL INFORMATION

Water may be discarded after it has been used. Nails and cups may be reused.

Energy Transfer and Specific Heat (cont.)

MISCONCEPTION ALERT

Students may predict that the final temperature of the combination of water and nails will be exactly halfway between the two initial temperatures. This prediction would be true only if the two substances had the same specific heat. You can explain that in Trial 1, for example, all (or almost all) of the energy lost from the warm nails was absorbed by the water, the temperature of the water did not change as much as the temperature of the nails because changing the temperature of water takes a greater amount of energy than changing the temperature of iron does. That is what specific heat measures.

Name _____ Class _____ Date _____

Skills Practice Lab

Energy Transfer and Specific Heat

Introduction

Heat is the energy transfer between objects that have different temperatures. Energy moves from objects that have higher temperatures to objects that have lower temperatures. If two objects are left in contact for a while, the warmer object will cool down and the cooler object will warm up until they eventually reach the same temperature. In this activity, you will combine equal masses of iron nails and water that have different temperatures to determine whether the iron nails or the water has a greater effect on the final temperature.

OBJECTIVES

Predict the final temperature when equal amounts of iron and water that have different temperatures are combined.

Measure the initial temperature of the iron, the initial temperature of the water, and the final temperature of the combined iron and water.

Compare the results of the experiment to the prediction.

MATERIALS

graduated cylinder, 100 mL   rubber band
marker   string, 30 cm long
metric balance   thermometer
nails (10–12)   water, cold
paper towels   water, hot
plastic-foam cups, 9 oz (2)

SAFETY

• Secure loose clothing and remove dangling jewelry. Don't wear open-toed shoes or sandals in the lab.

• In order to avoid burns, wear heat-resistant gloves when instructed to do so.

• If you are unsure of whether an object is hot, do not touch it.

• Check the condition of glassware before and after using it. Inform your teacher of any broken, chipped, or cracked glassware because it should not be used.

• Do not pick up broken glass with your bare hands. Place broken glass in a specially designated disposal container.

Name _____ Class _____ Date _____

Energy Transfer and Specific Heat (cont.)

Procedure

MAKE A PREDICTION

1. When you combine equal amounts of iron and water, each of which has a different temperature, will the final temperature be closer to the initial temperature of the iron, closer to the initial temperature of the water, or halfway in between?

CONDUCT AN EXPERIMENT

2. Use a rubber band to bundle the nails together. Measure and record the mass of the bundle in a data table like the one below. Tie a length of string around the bundle, leaving one end of the string 15 cm long. Use the marker to label the cups "A" and "B".

3. Put the bundle of nails into cup A. Let the end of the string hang over the side of the cup. Fill the cup with enough hot water to cover the nails, and set the cup aside for at least 5 min.

4. Use the graduated cylinder to measure the amount of cold water that has a mass equal to the mass of the nails (1 mL of water has a mass of 1 g). Record this volume in your data table. Pour the water from the graduated cylinder into cup B.

5. Measure and record the temperature of hot water that covers the nails in cup A and the temperature of the water in cup B. The temperature of the water in cup A represents the temperature of the nails.

6. Use the string to transfer the bundle of nails to the water in cup B. Use the thermometer to monitor the temperature of the water in cup B. When the temperature in cup B stops changing, record this temperature in the "Final temperature" column of your data table.

7. For Trial 2, repeat steps 3–6, but this time use cold water in step 3 and hot water in step 4. Start step 3 by filling cup A with cold water. In step 4, measure hot water to pour into cup B. Record all your measurements.

8. Empty the cups and dry the nails.

DATA TABLE

Trial	Mass of nails (g)	Volume of water (mL)	Initial temp of water and nails (°C)	Initial temp of water without nails (°C)	Final temp of water and nails combined (°C)
1					
2					

Name _____ Class _____ Date _____

Energy Transfer and Specific Heat (cont.)

1. Examining data In Trial 1, you used equal masses of cold water and nails. Did the final temperature support your prediction? Explain.

Answers will vary depending on the initial prediction.

_____

_____

2. Examining data In Trial 2, you used equal masses of hot water and nails. Did the final temperature support your prediction? Explain.

Answers will vary depending on the initial prediction.

_____

_____

3. Analyzing data In Trial 1 and Trial 2, which material—the water or the nails—changed temperature the most after you transferred the nails? Explain your answers.

The nails changed temperature more than the water in both trials.

Explanations should include references to initial and final temperatures.

_____

CONCLUSIONS

1. Drawing conclusions In Trial 1, the cold water gained energy. Where did the energy come from?

The energy came from the heated nails.

_____

_____

2. Drawing conclusions How does the energy gained by the nails in Trial 2 compare with the energy lost by the hot water in Trial 2? Explain.

The energy gained by the nails should be about the same as the energy lost

by the hot water (there might be some difference due to energy transfer to

the cup and air). The energy changes are the same because energy is con-

served. Any energy gained by the nails must come from somewhere, in this

case from the water.

Name _____ Class _____ Date _____

Energy Transfer and Specific Heat (cont.)

3. Drawing conclusions Which material seems to be able to hold energy better? Explain your answer.

The water appears to hold energy better. Students' explanations should

include that the temperature of the water changed less than the temperature

of an equal mass of iron.

4. Drawing conclusions Specific heat is a property of matter that tells how much energy is required to change the temperature of 1 kg of a material by 1 °C. Which material, iron or water, used in this activity has a higher specific heat (changes temperature less for a given amount of energy)?

Water has a higher specific heat capacity.

_____

_____

5. Applying conclusions Would it be better to have pots and pans made from a material with a high specific heat or a low specific heat? Explain your answer. (Hint: Do you want the pan or the food in the pan to absorb all the energy from the stove?)

Pots and pans should be made from a material with a low specific heat

capacity so that more energy from the stove will be transferred to the food

than to the pots and pans.

6. Evaluating results Share your results with your classmates. Discuss how you would change your prediction to include your knowledge of specific heat.

Accept all reasonable answers. Sample revised prediction: The final tempera-

ture would be closer to the original temperature of the substance with the

higher specific heat capacity.

EXTENSIONS

1. Designing experiments Design an experiment to compare the specific heat of nails made of iron with the specific heat of nails made of a metallic alloy.

A sample experiment could involve placing equal masses of iron nails and

alloy nails in hot water and waiting until they reach the same temperature.

Then the two sets of nails could be placed in separate containers of cold

water, each at the same temperature and with the same volume of water. The

nails that resulted in a higher final temperature would have the higher

specific heat.

Teacher's Notes and Answers

Boyle's Law

TIME REQUIRED
1 lab period

SKILLS ACQUIRED
Collecting data
Experimenting
Interpreting
Measuring
Organizing and Analyzing data

THE SCIENTIFIC METHOD

- **Make Observations:** In Procedure steps 2–5, students will observe how the volume of gas is affected by increasing pressure.

- **Analyze the Results:** In Analysis questions 1–4, students will analyze the data and results from their experiments.

- **Draw Conclusions:** In Conclusions questions 1–5, students will use their results to draw conclusions about the relationship between the pressure and volume of a gas.

- **Communicate the Results:** In In Conclusions questions 1–5, students will use their graphs to visually communicate the results of their experiments.

Teacher's Notes
MATERIALS

Check each apparatus for possible leaks. If the apparatus leaks, the piston will not return to its original position regardless of the number of times you twist the head. The most likely source of leakage is the bottom end of the syringe, where the needle is normally attached.

Each mass should be approximately 500 g. Less mass provides too small a change in volume. Greater mass may cause the pressure to be so excessive that air will begin to escape around the gasket. Masses of 500 g are ideal. All masses used by one lab group should be the same.

SAFETY CAUTIONS

Students should be cautious while working with the apparatus and when moving around the lab. The air in the syringe is under pressure. Falling masses can cause injury.

DISPOSAL INFORMATION

All materials used in this lab, including syringes, can be reused.

TIPS AND TRICKS

You may wish to have students calculate the uncertainty of the volumes and plot these on the graph of volume versus pressure. These steps could lead to a discussion about uncertainty and experimental error. In addition, it may help students appreciate why a line which does not actually go through every point is nevertheless an accurate representation of the data.

TECHNIQUES TO DEMONSTRATE

Show students how to give the piston several twists each time they change the pressure. The frictional force is significant, but if the piston is twisted until a stable volume is reached, reliable data should be obtained. Show students how to carefully stack the masses so that the pile is stable.

Name _____ Class _____ Date _____

Skills Practice Lab

Boyle's Law

Introduction

According to Boyle's law, all gases behave the same when compressed; that is, as increasing pressure is applied to a gas in a closed container, the volume of the gas decreases. Boyle's law may be stated mathematically as $P \sim \dfrac{1}{V}$ or $PV = k$

(where k is a constant). Notice that for Boyle's law to apply, two variables that affect gas behavior must be held constant: the amount of gas and the temperature.

In this experiment, you will vary the pressure of air contained in a syringe and measure the corresponding change in volume. Because it is often impossible to determine relationships by just looking at data in a table, you will plot graphs of your results to see how the variables are related. You will make two graphs, one of volume versus pressure and another of the inverse of volume versus pressure. From your graphs you can infer the mathematical relationship between pressure and volume in order to verify Boyle's law.

OBJECTIVES

Determine the volume of a gas in a container under various pressures.

Graph pressure-volume data to discover how the variables are related.

Interpret graphs, and verify Boyle's law.

MATERIALS

Boyle's law apparatus
carpet thread

objects of equal mass, approximately
500 g each (4)

SAFETY

- Secure loose clothing and remove dangling jewelry. Don't wear open-toed shoes or sandals in the lab.

- Wear safety goggles when working around chemicals, acids, bases, flames, or heating devices. Contents under pressure may become projectiles and cause serious injury.

- If any substance gets in your eyes, notify your instructor immediately and flush your eyes with running water for at least 15 minutes.

Name _____ Class _____ Date _____

Boyle's Law (cont.)

Procedure

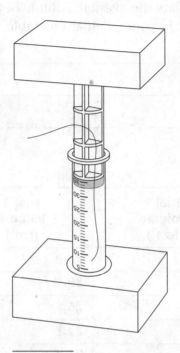

FIGURE A

1. Adjust the piston head of the Boyle's law apparatus so that it reads between 30 and 35 cm³. To adjust, pull the piston head all the way out of the syringe, insert a piece of carpet thread into the barrel, and position the piston head at the desired location, as shown in the figure above.

   Note: Depending upon the Boyle's law apparatus that you use, you may find the volume scale on the syringe abbreviated in cc or cm³. Both abbreviations stand for cubic centimeters (1 cubic centimeter is equal to one milliliter). The apparatus shown in the figure is marked in cm³.

2. While holding the piston in place, carefully remove the thread. Twist the piston several times to allow the head to overcome any frictional forces. Read the volume to the nearest 0.1 cm³. Record this value in your data table as the initial volume for zero masses.

3. Place one of the masses on the piston. Give the piston several twists to overcome any frictional forces. When the piston comes to rest, read and record the volume to the nearest 0.1 cm³.

4. Repeat step 3 for two, three, and four masses, and record your results.

5. Repeat steps 3 and 4 for at least two more trials. Record your results.

6. Clean all apparatus and your lab station at the end of this experiment. Return equipment to its proper place. Ask your teacher how to dispose of any waste materials.

Name _____ Class _____ Date _____

| Boyle's Law (cont.)

ANALYSIS

1. Organizing data Calculate the average volume of the three trials for masses 0–4. Record your results in your calculations table.

Answers will vary.

_____

_____

_____

DATA TABLE

Pressure (number of weights)	Trial 1 Volume (cm³)	Trial 2 Volume (cm³)	Trial 3 Volume (cm³)
0	33.0	33.0	33.0
1	29.7	29.5	29.0
2	26.8	26.8	26.0
3	24.3	24.1	23.5
4	22.0	21.8	21.2

2. Organizing data Calculate the inverse for each of the average volumes. For example, if the average volume for three masses is 26.5 cm³, then 1/volume = $1/26.5/cm^3 = 0.0377/cm^3$.

Answers will vary.

_____

_____

_____

CALCULATIONS TABLE

Pressure (number of weights)	Average Volume (cm³)	1/Volume ($\times 10^2$/cm³)
0	33.0	3.03
1	29.4	3.40
2	26.5	3.77
3	24.0	4.17
4	21.7	4.61

Name _____ Class _____ Date _____

Boyle's Law (cont.)

3. Constructing graphs Plot a graph of volume versus pressure. Because the number of masses added to the piston is directly proportional to the pressure applied to the gas, you can use the number of masses to represent the pressure. Notice that pressure is plotted on the horizontal axis and volume is plotted on the vertical axis. Draw the smoothest curve that goes through most of the points.

Graphs will vary; graphs should *not* show a linear relationship between

volume and pressure.

_____

Volume Versus Pressure

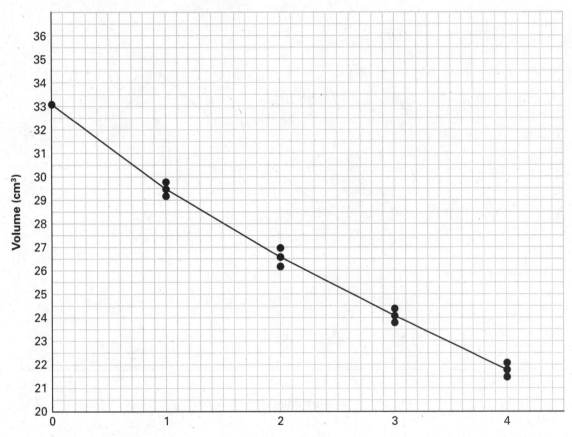

Name _____ Class _____ Date _____

Boyle's Law (cont.)

4. Constructing graphs Plot a graph of 1/volume versus pressure. Notice that pressure is on the horizontal axis and 1/volume is on the vertical axis. Draw the best line that goes through the majority of the points.

Graphs will vary; this graph should show a roughly linear relationship

between inverse volume and pressure.

_____

1/V Versus Pressure

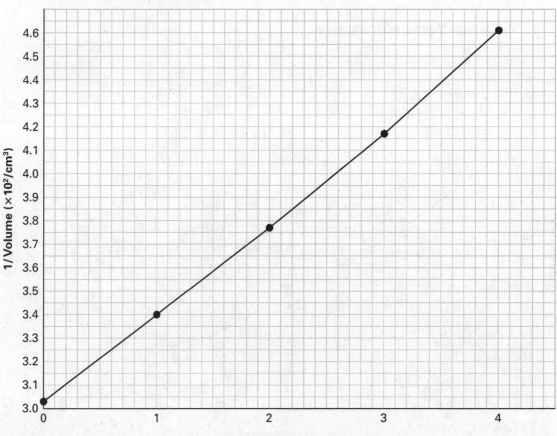

CONCLUSIONS

1. Analyzing graphs Do your graphs indicate that a change in volume is directly proportional to a change in pressure? Explain your answer.

no; The graph is not a straight line. Furthermore, as the pressure increases,

the volume decreases.

_____

Name _____ Class _____ Date _____

Boyle's Law (cont.)

2. Analyzing graphs Based on your graphs, what do you conclude about the mathematical relationship between the pressure applied to a gas and the corresponding volume of the gas?

The graph of 1/volume versus pressure is approximately a straight line

(within experimental uncertainty). Therefore, pressure must be proportional

to 1/volume. Pressure and volume are inversely related.

3. Making predictions Use your graph to predict the volume of the gas if 2.5 masses were used.

Answers will vary but should lie between the volume with two masses and

the volume with three masses.

4. Defending conclusions Use the kinetic theory description of a gas to explain the observed relationship between the pressure and volume of a gas.

Gas pressure results from particles hitting the sides of the container. If the

size of the container is decreased while the number of molecules remains the

same, the number of hits is increased.

5. Applying conclusions If a normal sea-level recipe is used to prepare a cake at a location 1000 meters below sea level, the cake will be much flatter than expected. Explain why, and offer a solution. (Hint: Consider how barometric pressure differs at this altitude and what effect that may have on the ability of the cake to rise.)

At the lower altitude, the barometric pressure is greater. As a result, Boyle's

law predicts that the gas bubbles will be smaller. To compensate for the

reduced volume of gas, one should add more baking powder or less flour to

the batter.

Name _____ Class _____ Date _____

Boyle's Law (cont.)

Extensions

1. **Research and communications** Use the Internet or a library to look up
   another fundamental gas law, such as Charles's law or Avogadro's law. Write
   down the mathematical expression of the law. State which quantities the law
   assumes to be held constant and which quantities may change.

   **Charles's law states that if pressure and amount remain constant, the ratio
   of volume to temperature is a constant ($V/T = k$). Avogadro's Law states
   that if temperature and pressure remain constant, the ratio of volume to
   amount stays the same ($V/n = k$).**

       T52

Teacher's Notes and Answers

Creating and Measuring Standing Waves

TIME REQUIRED

1 lab period

SKILLS ACQUIRED

Collecting data
Recognizing patterns
Measuring
Organizing and analyzing data
Predicting

THE SCIENTIFIC METHOD

- **Make Observations:** In the Procedure, students create standing waves and observe the properties of those standing waves.

- **Analyze the Results:** In the Analysis, students calculate wavelengths and wave speeds and look for patterns in their data and results.

- **Draw Conclusions:** In the Conclusions, students draw general conclusions about how the variables in the wave equation are related, and they make predictions based on their conclusions.

- **Communicate Results:** In the Analysis and Conclusions, students communicate their results by providing written answers to the questions.

Teacher's Notes

MATERIALS

The Extension exercise asks students to design an experiment to test wave speed by using different kinds of rope. You may have individual groups carry out this experiment, or you can get similar results by providing different groups with different kinds of rope and then comparing wave speeds from the different groups at the end. If you want to keep things simple, provide each group with the same kind and same length of rope. All of the wave speeds should then be about the same.

SAFETY CAUTIONS

Warn students to clear the area and to keep hands and clothing out of the way of the ropes while the ropes are in motion.

TECHNIQUES TO DEMONSTRATE

You may wish to demonstrate how to create standing waves on a rope. Focus in particular on keeping the motion of the rope in a vertical plane and minimizing the motion of the hand. Explain what is meant by a "loop" (each loop corresponds to an antinode).

Name _____ Class _____ Date _____

Skills Practice Lab
Creating and Measuring Standing Waves

Introduction

When you fix a rope at one end and then shake the other end up and down, waves travel down the rope. You can vary the frequency and wavelength of the waves by changing the rate at which you shake the rope. If the waves are of certain wavelengths, the waves will appear to stop traveling along the rope, and certain points on the rope, called *nodes*, will be completely motionless. Waves that behave this way are called *standing waves*. In this experiment, you will create several different standing waves on a rope. You will calculate the frequency, wavelength, and wave speed for each standing wave and then will look for patterns in these values.

OBJECTIVES

Create standing waves on a rope fixed at one end.

Calculate frequency, wavelength, and wave speed of standing waves.

Compare frequency, wavelength, and wave speed of different standing waves.

MATERIALS

clothesline rope, 3 m long
meterstick
stopwatch

Safety

• Secure loose clothing and remove dangling jewelry. Don't wear open-toed shoes or sandals in the lab.

Name _____ Class _____ Date _____

Creating and Measuring Standing Waves (cont.)

Procedure
PULLING MASSES

1. Tie one end of the rope to a fixed support, such as a doorknob. Hold the free end of the rope, and measure the length of the rope from your hand to the knot at the fixed end. Record the length in a data table like the one shown below.

2. Shake the free end of the rope up and down, starting slowly. Be sure not to move the rope side to side or in a circle. Adjust the frequency until the entire rope is vibrating up and down as a whole (making one "loop," even though the loop is confined to a plane). Try to keep the rope moving like this while minimizing the motion of your hand. There should now be a node at each end of the rope and an antinode in the middle.

3. Using a stopwatch, time how long it takes to shake the rope up and down through 10 cycles (one up-and-down motion is one cycle). Divide this total time by 10 to calculate the frequency of the standing wave. Record the frequency in your data table.

4. Increase the frequency by shaking the free end of the rope up and down faster. Continue to increase the frequency until there is a node in the middle of the rope. Now the rope should have two "loops," with two antinodes.

5. Calculate the frequency of this standing wave with two loops by using the same method as in step 3. Record the frequency in your data table.

6. Increase the frequency again until you have a standing wave with three antinodes (three loops). Determine the frequency as in step 3, and record the frequency in your data table.

DATA TABLE

Total length of rope: _____ m			
Number of antinodes	Frequency (Hz)	Wavelength (m)	Wave speed (m/s)
1			
2			
3			

Name _____ Class _____ Date _____

| *Creating and Measuring Standing Waves (cont.)*

ANALYSIS

1. **Organizing data** Nodes in a standing wave are always one-half of a wave-length apart. Therefore, with one loop in the rope, the rope is one-half of a wavelength long. Calculate the total wavelength by multiplying the length of the rope by 2. Record this wavelength in your data table.

   The wavelength for a standing wave with one antinode should be twice the

   length of the rope.

2. **Organizing data** With two loops in the rope, the rope is one wavelength long. With three loops, the rope is one-and-a-half wavelengths long. Calculate the wavelength for these two cases, and record the wavelengths in your data table.

   The wavelength for a standing wave with two antinodes should equal the

   length of the rope. The wavelength for a standing wave with three antinodes

   should be $^2/_3$ times the length of the rope.

3. **Recognizing patterns** As the frequency increased, did the wavelength increase or decrease?

   As frequency increases, wavelength decreases.

4. **Organizing data** Once you know the frequency and wavelength of a wave, you can calculate the speed of the wave by using the wave-speed equation:

   $$wave\ speed = frequency \times wavelength.$$

Calculate the wave speed for each of the three standing waves that you produced.

   Wave speeds will vary from group to group, but should equal *frequency* ×

   ***wavelength* for each standing wave. The three wave speeds calculated by a**

   single group should be approximately equal.

Name _____ Class _____ Date _____

Creating and Measuring Standing Waves (cont.)

CONCLUSIONS

1. Drawing conclusions How are the wave speeds of the three different standing waves related?

The wave speeds are approximately equal.

_____

_____

_____

2. Drawing conclusions How are the frequencies of the three different standing waves related?

The frequencies of the standing waves are related by simple whole-number

ratios. The frequency of the standing wave that has two antinodes is twice

the frequency of the standing wave that has one antinode. The frequency of

the standing wave that has three antinodes is three times the frequency of

the standing wave that has one antinode.

3. Making predictions What would the frequency, wavelength, and wave speed of a standing wave with four antinodes on your rope be?

Answers may vary depending on the data. The frequency will be four times

the frequency of the standing wave that has one antinode. The wavelength

will be one-fourth of the total length of the rope. The wave speed will

approximately equal the wave speeds calculated for the other cases.

4. Making predictions How could you change the experimental setup to produce standing waves that have a different wave speed?

You could change the wave speed by using a different kind of rope.

_____

_____

_____

Name _____ Class _____ Date _____

Creating and Measuring Standing Waves (cont.)

EXTENSIONS

1. **Designing experiments** Design an experiment to determine how the speed of standing waves is related to the density (mass per unit length) of the rope used. The results of the experiment should include a graph. If you have time and your teacher approves, carry out your experiment by using at least three kinds of rope.

Experiments should involve creating standing waves with at least three kinds of rope. The mass per unit length of each kind of rope can be determined by weighing the rope and dividing the mass by the total length. With each rope, frequency and wavelength should be determined for at least one standing wave by using methods similar to the procedure of the experiment that they already carried out. The frequency and wavelength can then be used to calculate wave speed. A graph can then be plotted to show wave speed as a function of mass per unit length.

Teacher's Notes and Answers

Mirror Images

TIME REQUIRED
1 lab period

SKILLS ACQUIRED
Classifying
Collecting data
Constructing models
Identifying/Recognizing patterns
Measuring
Organizing and Analyzing data

THE SCIENTIFIC METHOD

- **Make Observations:** In the procedure steps, students will make observations regarding virtual images, flat mirrors, and curved mirrors.

- **Analyze the Results:** In Analysis questions 1–7, students will analyze the data and results from their experiments.

- **Draw Conclusions:** In Conclusions questions 1–4, students use their results to draw conclusions about mirror images.

Teacher's Notes
MATERIALS

If a reverse eye chart is unavailable, use the normal eye chart for both parts of the activity.

Any small object can substitute for the T-pin and pencil eraser.

Mount all mirrors in cardboard to protect students from sharp edges. Use a craft knife to cut cardboard squares to make a frame for each mirror, and tape the mirror securely into the frame.

For the curved mirrors, you may either use separate concave and convex mirrors, or a double-sided curved mirror (one concave side and one convex side).

SAFETY CAUTIONS

Secure mirrors to the wall with strong tape and to the table with clamps or strong tape. The mirror taped to the wall should be lightweight and no larger than 10 cm by 15 cm.

The pencil eraser used on the T-pin should be new and not worn. The purpose of putting the eraser on the tip of the pin is to protect the student's eyes (it also makes the pin easier to see). Make sure that students wear safety goggles. It is possible to substitute another small object, such as a larger eraser or the cap of a pen, for the T-pin.

MISCONCEPTION ALERT

Students may confuse the terms *concave* and *convex*. You may point out to them that a concave mirror is recessed inward, like a cave.

TIPS AND TRICKS
Classroom Organization

Students can work alone or in groups of two or more.

It is more efficient to set up the lab in stations so that groups can rotate from station to station.

Checkpoints

Step 3: For the bench test of this lab, line 11 on the eye chart became blurry at a distance of 3.6 m.

Step 4: Make sure that the mirror is small and lightweight. A mirror that is too heavy will fall and break. Tape all sides of the mirror to the wall.

Step 6: For the bench test, line 11 became unreadable at 1.75 m.

Step 8: Secure mounted mirrors to the table by using clamps or strong tape to prevent broken-glass hazard.

Step 13: The angle between the incoming beam and the nearest perpendicular line should be approximately equal to the angle between the outgoing beam and the nearest perpendicular line.

Step 18: In the bench test at close range (20 cm), objects appear enlarged and upright; at middle range (30 cm), objects appear reduced and upright; at long range (65 cm), objects appear greatly reduced and upright.

Step 21: In the bench test at close range (20 cm), objects appear enlarged and upright; at middle range (30 cm), objects appear greatly enlarged and upright; at long range (65 cm), objects appear reduced and inverted.

TECHNIQUES TO DEMONSTRATE

Show students how to properly secure mirrors to the table by using mirror supports and to the wall by using strong tape.

Name _____ Class _____ Date _____

Skills Practice Lab

Mirror Images

Introduction

Mirrors form images by reflecting light coming from objects. When you view an image in a mirror with your eyes, you are seeing a virtual image, which usually appears to exist somewhere behind the mirror. Depending on the shape of the surface of the mirror, this mirror image may be reversed left to right, turned upside down, or magnified to look larger or smaller. In this lab, you will observe images in a flat mirror, a convex mirror (curved outward), and a concave mirror (curved inward). You will also measure various distances and angles to learn more about the properties of the images formed by these different types of mirrors.

OBJECTIVES

Describe the images of objects in various kinds of mirrors.

Draw diagrams showing paths of light rays reflected from mirrors.

Compare tlight path lengths and angles of incidence and reflection.

MATERIALS

eye charts, both normal and reverse	pencil
meterstick	protractor
mirror, curved	ruler or straightedge
mirror, small and flat	tape
mirror supports	T-pin with pencil eraser
paper	white paper

SAFETY

- Secure loose clothing and remove dangling jewelry. Don't wear open-toed shoes or sandals in the lab.

- Never look directly at the sun through any optical device or use direct sunlight to illuminate a microscope.

- If any substance gets in your eyes, notify your instructor immediately and flush your eyes with running water for at least 15 minutes.

- Check the condition of glassware before and after using it. Inform your teacher of any broken, chipped, or cracked glassware because it should not be used.

- Do not pick up broken glass with your bare hands. Place broken glass in a specially designated disposal container.

Name _____ Class _____ Date _____

| Mirror Images (cont.)

Procedure

VIRTUAL IMAGES

1. Secure the normal eye chart to the wall by using strong tape.

2. Choose any line on the chart, and step back until the line can no longer be read clearly. Use masking tape to mark the position on the floor where you are standing. Label this position "reading point."

3. Measure the distance from the eye chart to the reading point with a meterstick. Record this distance in your notebook, using the appropriate SI units (meters). Also, record the number of the line that you were trying to read.

4. Secure a small, flat mirror against the wall at chest level by using strong tape.

5. Place the back of the reverse eye chart against your chest. Position the chart so that the line that you read appears in the mirror. Step back from the mirror, and hold the eye chart against your chest until the image of this line is barely readable.

6. Use masking tape to mark the position on the floor where you are standing. Label this position "new point."

7. Measure the distance from the eye chart to the new point. Record this distance in your notebook, using the appropriate SI units.

FLAT MIRRORS

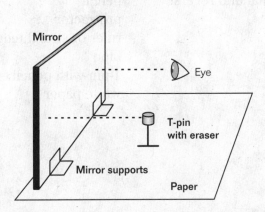

8. Using two mirror supports, vertically stand one flat mirror on a table, away from the edge, as shown in the figure above. Place a sheet of white paper on the tabletop so that the front of the mirror faces the paper. Tape the paper and mirror supports to the table so that they do not slide.

9. Remove the eraser from a pencil. Secure the eraser on the tip of the T-pin to cover the point. Using tape, carefully secure the T-pin on the tabletop, with the T side down, in front of the mirror. The point with the eraser should be pointing up.

10. Wearing a pair of safety goggles, move your head to one side of the pin. Close one eye, and place your open eye at the level of the tabletop. Observe the image of the pin in the mirror.

Name _____ Class _____ Date _____

Mirror Images (cont.)

11. Use a ruler to draw a straight line on the paper from the image of the pin in the mirror to the position of your eye. Label this line "outgoing beam." Use a ruler to draw a straight line from the object to the mirror's surface. Label this line "incoming beam." Both lines should meet at the same point on the mirror's surface.

12. Draw a line on the paper from the position of your eye perpendicular to the mirror's surface. Draw a line from the object perpendicular to the mirror's surface. Both lines should be parallel to each other. These lines will form angles with the lines you drew in step 11.

13. Measure the angle between the line labeled "outgoing beam" and the nearest perpendicular line. Measure the angle between the line labeled "incoming beam" and the nearest perpendicular line. Record these angles in your notebook, using units of degrees.

14. Move your eye to a new position. Repeat steps 10–13.

15. Move your eye to a third position. Repeat steps 10–13.

CURVED MIRRORS

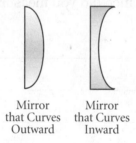

Mirror that Curves Outward   Mirror that Curves Inward

16. Obtain a curved mirror. Use one mirror support to hold the mirror upright on the bench. Place the mirror so that you are facing the side that curves outward (the convex side).

17. Place an object at various distances from the mirror. Look at the image of the object in the mirror.

18. Observe and record in your notebook how the image appears. Include the object's position (close to the mirror or far from the mirror), the size of the image (enlarged or small), and the orientation of the image (upright or upside down).

19. Turn the mirror around so that you are facing the side that curves inward (the concave side).

20. Place an object at various distances from the mirror. Look at the image of the object in the mirror.

21. Observe and record in your notebook how the image appears. Include the object's position (close to the mirror or far from the mirror), the size of the image (enlarged or small), and the orientation of the image (upright or upside down).

Name _____ Class _____ Date _____

| Mirror Images (cont.)

Analysis
VIRTUAL IMAGES

1. Describing events Describe the image of the reverse eye chart you saw on the surface of the mirror. Compare it with the appearance of the normal eye chart.

The image of the reverse eye chart seen in the mirror is identical to the

appearance of the normal eye chart viewed directly (it is no longer

reversed).

2. Examining data What distance did you measure between the mirror and the reverse eye chart?

Answers will vary.

3. Examining data What distance did you measure between the starting point and the eye chart on the wall?

Answers will vary.

FLAT MIRRORS

4. Organizing data In your notebook, draw the experimental setup as viewed from above. Include the lines and angles for each trial.

Diagrams should show that the incoming and outgoing angles are equal in

each trial.

CURVED MIRRORS

5. Describing events How did the image appear when the object was in front of the convex mirror?

When an object is in front of a convex mirror, its image appears upright and

smaller than the object.

Name _____ Class _____ Date _____

Mirror Images (cont.)

6. Describing events How did the image appear when the object was close to the concave mirror?

When an object is close to a concave mirror, its image appears upright and

larger than the object.

7. Describing events How did the image appear when the object was far away from the concave mirror?

When an object is far from a concave mirror, its image appears inverted and

smaller than the object.

Conclusions

1. Drawing conclusions Compare your answers to Analysis questions 2 and 3 above. What is the relationship between the distances?

The distance from the chart to the mirror is approximately one-half the

distance from the student to the chart. Some students may realize that the

distance between the image and the eye is actually the same in both cases.

2. Drawing conclusions Compare the two angles measured in step 13 for each position. What is the relationship between the angles?

The angles are equal.

3. Classifying Which kinds of mirrors used in this experiment are capable of producing images that are reversed left to right?

all three types: flat, concave, and convex

4. Classifying Which kinds of mirrors used in this experiment are capable of producing images that are inverted (upside down)?

Only the concave mirror can create an inverted image.

Name _____ Class _____ Date _____

Mirror Images (cont.)

Extensions

1. **Research and communications** Many side mirrors on automobiles have a warning label, "Objects in mirror are closer than they appear." What kind of mirror do you think these are? What might be the advantage of using this kind of mirror for this application? Research these types of mirrors on the Internet or in a library to check your answers.

   Automobile side mirrors that have the warning label are convex mirrors. The primary advantage of using a convex mirror is that it provides a wider field of view.

2. **Building models** If you were designing a house of mirrors for a carnival, what kind of mirror would you use to create an upside down image? Would the person looking in the mirror have to be close to the mirror or far away from it to see his or her image upside down?

   You would use a concave mirror. The person would have to stand far from the mirror.

Teacher's Notes and Answers

Converting Wind Energy into Electricity

TIME REQUIRED

1 lab period

SKILLS ACQUIRED

Constructing models
Experimenting
Predicting
Recognizing patterns

THE SCIENTIFIC METHOD

- **Make observations:** In the Procedure, students observe how changing parts of the system affects the system's operation and energy output.

- **Analyze the results:** In the Analysis, students look for patterns in their observations and offer explanations.

- **Draw conclusions:** In the Conclusions, students draw general conclusions about how the system should be set up to work best and predict how changes to the system would affect operation.

- **Communicate results:** In the Analysis and Conclusions, students communicate results by providing written answers to the questions.

Teacher's Notes
MATERIALS

The pencils used to make the pinwheels should be round and smooth, not hexagonal. The pencils should be unsharpened (completely flat on the end opposite the eraser) and long enough to pass through the shoe boxes with at least 4 cm to spare.

You may use wooden dowels instead of pencils. In that case, you should use a thumbtack to carefully poke a hole in one end of each dowel before class. Do not leave this step for the students to do; they could easily poke or cut themselves. One advantage of using dowels is that they can be cut to any length so that they pass through the shoe boxes with ample length to spare.

SAFETY INFORMATION

Advise students to use caution when using the blow dryers. Blow dryers can become very hot when used for extended periods. Students should avoid touching the tip of the dryer to clothing, paper, or the turbines. Because blow dryers may blow dust or small objects, students should wear safety goggles during this experiment.

Warn students that wire coils may become hot. (Because the current is low, the coils probably will not get very hot in this experiment, but it is a good message to convey for future experiments.)

Warn students about handling sharp objects, such as scissors and thumbtacks.

TECHNIQUES TO DEMONSTRATE

Show students a sample turbine so that they will know how to fold the poster board correctly.

If the students have not used voltmeters before, show them how to turn on a voltmeter and how to set the meter to the proper voltage range.

TIPS AND TRICKS

Balancing the turbine and shaft properly (step 11) is probably the most difficult—and most important—part of the Procedure. Help students who are having trouble with this step. The key is to have weight evenly distributed around the shaft and to have the magnet centered on the end of the shaft. Also, make sure that the edges of the holes through the box are smooth.

Name _____ Class _____ Date _____

Skills Practice Lab

Converting Wind Energy into Electricity

Introduction

Electricity is a form of energy. Electricity can come from many different sources. The electricity in your house comes from a power plant, which may generate electricity by harnessing energy from burning coal or natural gas, nuclear reactions, flowing water, or wind. The energy from any one of these sources can be converted into electrical energy, which can in turn be converted into other forms of energy. For example, an electric stove or a toaster turns electrical energy into heat that can be used for cooking. A lamp turns electrical energy into light.

In this lab, you will build a device to convert the energy in wind into electrical energy. The electrical energy will then be converted into light as the electricity powers a light-emitting diode (LED).

OBJECTIVES

Construct a device that uses wind energy to generate electricity.

Use the device to power a light-emitting diode (LED).

Describe how changing various parts of the system affects the generation of electricity.

MATERIALS

alligator clips and wire (2)
ALNICO cylindrical cow magnet hole punch
blow dryer, 1500 W, 60 Hz
LED, 2.0 V
metric ruler
modeling clay
pencil, round and unsharpened

pipe cleaners (2)
poster board, 10 cm × 10 cm
PVC wire, 100 ft coil
scissors
shoe box
thumbtack
transparent tape
voltmeter or multimeter

Safety

- Do not place electrical cords in walking areas or let cords hang over a table edge in a way that could cause equipment to fall if the cord is accidentally pulled.

- Be sure that equipment is in the "off" position before plugging it in.

- Be sure to turn off and unplug electrical equipment when finished.

- If you are unsure of whether an object is hot, do not touch it.

- Use knives and other sharp instruments with extreme care.

Name _____ Class _____ Date _____

Converting Wind Energy into Electricity (cont.)

- Never cut objects while holding them in your hands. Place objects on a suitable work surface for cutting.

Procedure
BUILDING A WINDMILL

1. Use the ruler as a straight edge to draw two diagonal lines on the poster board. Each line should go from one corner to the opposite corner. Punch a hole in the center of the square (where the lines cross). Make a 5 cm cut from each corner of the square toward the center (along the lines).

2. Slide the eraser end of the pencil through the hole in the center of the square so that the eraser end of the pencil sticks out about 3 cm. Carefully bend one corner of the square toward the eraser. Be careful not to make a fold in the poster board. Then, work your way around the square by bending alternating corners in to the eraser until the poster board is in a pinwheel shape, as shown in **Figure 1.** Use a thumbtack to pin the four folded corners to the eraser. This step can be tricky, so be patient and be careful not to poke yourself with the tack. When you are finished, you should have a turbine on a shaft.

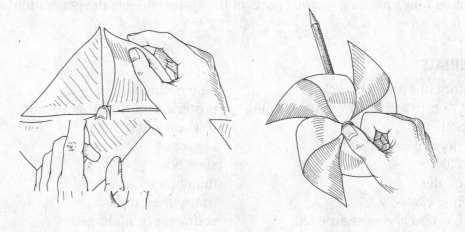

FIGURE 1

3. Push the back end of the turbine away from the eraser to create a lot of space in the turbine to catch wind. Securely tape the back end of the turbine to the pencil so that the shaft will turn when the turbine turns.

4. Place the box on the table so that the opening faces up. Carefully punch a hole in the long side of the shoebox, at least 8 cm above the table. Punch another hole on the other long side of the box so that the holes are opposite one another.

5. Use tape to attach the box to the table, or place a weight inside the box. Slide the shaft through the two holes so that the turbine is near the box and the other end of the shaft sticks out the other side of the box.

Name _____ Class _____ Date _____

Converting Wind Energy into Electricity (cont.)

6. Wrap pipe cleaners around the pencil where it exits the box on both sides. Tape the pipe cleaners to the pencil to prevent the shaft from sliding when wind blows on the turbine. Blow on the turbine to make sure that the turbine and shaft turn freely.

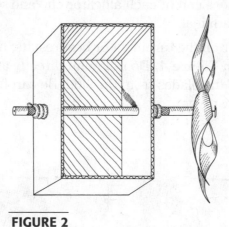

FIGURE 2

7. Tape the bar magnet to the end of the shaft opposite the turbine so that the shaft and magnet form a **T**, as shown in **Figure 3.** The two poles of the magnet should stick out to the left and right.

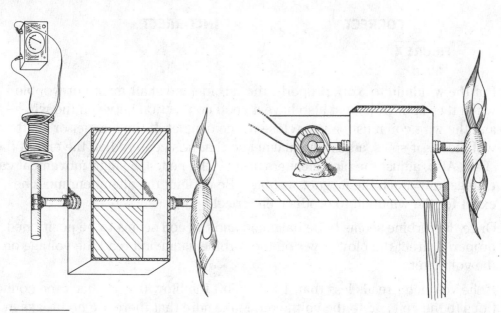

FIGURE 3

8. Place the coiled wire beside the magnet. Point the magnet's pole to the center of the coil. Place the coil on a base of modeling clay to hold the coil in position. Use the modeling clay to adjust the height of the coil so that the center of the coil is the same height as the magnet when it points to the coil. (The center of the coil should be at the same height as the shaft, about 8 cm above the table.)

Name _____ Class _____ Date _____

Converting Wind Energy into Electricity (cont.)

GENERATING ELECTRICITY

9. Find the two loose ends of the wire. One wire is inside the coil, and the other wire is outside the coil. Tape the outside of the coil so that it does not unravel. Connect each loose wire end to a separate alligator clip and wire assembly. Clip the loose end of each alligator clip and wire assembly to a different voltmeter terminal.

10. Direct the blow dryer on the turbine. For best results, hold the blow dryer at an angle, as shown in **Figure 4.** Do not point directly at the center of the turbine but at one of the blades from the side. Be careful not to touch the turbine with the blow dryer.

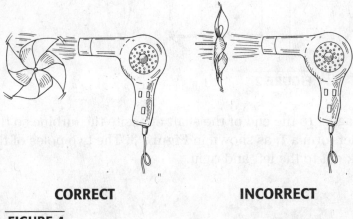

CORRECT   **INCORRECT**

FIGURE 4

11. For the windmill to work properly, the turbine and shaft must not wobble when they spin. You must also have a good connection between the wires, and the wire coil must be very close to the magnet. If the magnet or shaft wobbles as it spins, adjust the magnet so that it is centered on the end of the shaft. Also, adjust the pipe cleaners around the pencil; a slight movement can change how much the magnet wobbles. Be patient; the alignment must be exact for the windmill to produce electrical energy.

12. Once the turbine seems to be balanced and the coil seems to be positioned properly, direct the blow dryer on the turbine again and read the voltage on the voltmeter.

13. If the voltmeter reads less than 1 V, turn off the blow dryer. Check the connections to the coil and to the voltmeter. Make sure that there are no breaks in the insulation on the wire. Adjust the wire spool and the spinning magnet so that they are as close together as possible without touching. Direct the blow dryer on the turbine again. If the voltage is still less than 1 V, move the coil to slightly different positions while the turbine is spinning to try to get a higher voltage. Be careful not to let the coil touch the magnet while the magnet is spinning.

Name _____ Class _____ Date _____

Converting Wind Energy into Electricity (cont.)

USING ELECTRICITY TO LIGHT AN LED

14. Once the voltmeter consistently reads over 1 V, turn off the blow dryer. Disconnect the voltmeter, and clip the alligator clips and wire assembly to each wire of an LED.

15. Direct the blow dryer on the turbine again. You should see the LED glow.

16. Move the blow dryer so that it points to the turbine at different angles. Observe how the brightness of the LED is affected.

17. While the LED is glowing, move the coil around so that the shaft points straight toward the center of the shaft. Observe how the brightness of the LED is affected.

18. When you are finished, take apart your windmill and put all supplies back in their proper place.

ANALYSIS

1. Describing events Starting with the wind from the blow dryer, list as many forms of energy as you can in the system you have built. Your list should have at least three forms of energy.

Answers should include at least three of the following: wind energy (kinetic

energy) from the blow dryer; kinetic energy of the turbine, shaft, and/or

magnet; electromagnetic energy in the spinning magnet; electrical energy in

the coil and wires; and light (electromagnetic energy) from the LED.

2. Recognizing patterns What happened to the brightness of the LED when you moved the blow dryer away from the position shown in Figure 4?

In most cases, moving the blow dryer away from the position shown in the

figure causes the LED to dim. In some cases, moving the blow dryer may

cause the LED to glow brighter.

3. Explaining events Why did moving the blow dryer change the brightness of the LED?

In different positions, the blow dryer makes the turbine spin at different

speeds. The faster the turbine spins, the brighter the LED glows (because

there is more energy in the system).

Name _____ Class _____ Date _____

Converting Wind Energy into Electricity (cont.)

4. Describing events What happened to the brightness of the LED when you moved the coil in step 17?

The LED grew dimmer or stopped glowing.

_____

_____

_____

CONCLUSIONS

1. Drawing conclusions Based on your observations, in what position should the coil be relative to the spinning magnet in order to produce the most electricity?

To generate the most electricity, the coil should be close to the magnet and

perpendicular (at a right angle) to the plane of the magnet's rotation.

_____

_____

2. Making predictions Would the LED still glow if you turned the magnet so that it pointed straight out from the shaft?

no

_____

_____

_____

3. Drawing conclusions Based on your observations, in what position should the blow dryer be relative to the turbine in order to produce the most electricity?

Answers may vary. In most cases, students should find that the blow dryer

works best if it is close to the turbine and pointed toward the turbine at an

angle.

_____

Name _____ Class _____ Date _____

Converting Wind Energy into Electricity (cont.)

EXTENSIONS

1. **Building models** In a real-world situation, the wind blowing on a windmill does not always come from the same direction. How could you modify your windmill design to solve this problem?

   You could put the windmill on a rotating base so that the windmill can be

   turned toward the wind. You could also add a rudder to the back of the

   windmill so that the windmill would turn automatically.

   _____

   _____

2. **Building models** In a real-world situation, the wind does not always blow enough to turn a windmill. What could be added to a windmill system to ensure a steady supply of electricity when the wind stops blowing (at least for a while)?

   You could attach a rechargeable battery to the coil. When the wind blows,

   the battery charges. When the wind stops blowing, the energy stored in the

   battery can be used to generate electricity.

   _____

   _____

Teacher's Notes and Answers

Constructing and Using a Compass

TIME REQUIRED
1 or 2 lab periods

SKILLS ACQUIRED
Collecting data
Constructing models
Experimenting
Organizing and analyzing data
Predicting

THE SCIENTIFIC METHOD

- **Make Observations:** In steps 4 and 11 of the Procedure, students observe the behavior of a compass under different conditions.

- **Form a Hypothesis:** In steps 3 and 10 of the Procedure, students predict how the compass will behave.

- **Analyze the Results:** In the Analysis, students analyze their results.

- **Draw Conclusions:** In the Conclusions, students evaluate their results and make additional predictions.

- **Communicate the Results:** In the Analysis and Conclusions, students communicate results by providing written answers to the questions.

Teacher's Notes
MATERIALS

Steps 12–21 of the Procedure are optional. In these steps, students use the compasses to navigate a course and to locate objects that they have placed on the course. This part of the activity works best in a large outdoor area, away from traffic. If you want students to do this part of the Procedure, prepare as follows. Preparation could take up to 1 h.

ADDITIONAL MATERIALS FOR PREPARATION

10 index cards with names of different objects
2 additional index cards for each lab group
compass
masking tape or colored string
metric tape measure

First, measure a 10 m distance, and mark the beginning and ending points with tape or colored string. Students will use this distance to calibrate their paces to distance. Next, position five index cards containing the names of various objects, and labels naming the objects, around an open area. Mark a starting point, and then walk from that point to one of the objects while holding a compass. On another index card, record the compass direction, distance walked, and the name of the object on the card. From there, walk to a second card and repeat the process. Continue until you have a set of directions for all five objects. Repeat to generate a different set of directions for each group that will be doing the activity. When you have finished generating directions, place the five remaining objects in arbitrary positions. Make an index card that shows the directions that each group should follow. Each card should have a blank column that students will fill in with the names of the objects, as shown below. This will encourage students to navigate very carefully. You can verify that each group correctly navigated the course by comparing the objects in the "Object found" column to your master index cards.

SAMPLE NOTE CARD

Directions	Object found
1. Walk 1 m north.	
2. Walk 20 m east.	
3. Walk 5 m north.	
4. Walk 30 m west.	
5. Walk 5 m southeast.	
6. Return to start.	

SAFETY CAUTIONS

Warn students not to scrape or puncture themselves when stroking the paper clip with the magnet.

Follow your school's guidelines for outdoor activities if students will be doing the orienteering course.

DISPOSAL INFORMATION

Water should be disposed of in a sink. You may reuse all other materials.

TIPS AND TRICKS

This activity works best in groups of three or four students. Before you begin, have students put two bar magnets together to observe the attraction between opposite poles. Compare the Earth to a magnet—Earth is surrounded by a magnetic field and has north and south poles. Because Earth's magnetic field is used to define which pole of a compass is the "north-seeking pole," the north pole of the compass is by definition attracted to the magnetic north pole of Earth.

After students construct their compasses, demonstrate how to use a compass correctly. To encourage students to measure accurately, tell them that you have placed extra objects on the course.

Name _____ Class _____ Date _____

Skills Practice Lab

Constructing and Using a Compass

Introduction

Like a bar magnet, Earth has a magnetic field, with a magnetic north pole and a magnetic south pole. A compass takes advantage of Earth's magnetic field to orient a needle along the north-south direction. In a commercial compass, the north-pointing end of a compass needle usually has a colored arrow on it, so you can tell north from south. If you place a bar magnet near a compass, the north-pointing end of the needle will actually point toward the south pole of the magnet. This concept can be confusing—just remember that the North Pole of the Earth is like the south pole of a magnet.

In this activity, you will use ordinary objects—a paper clip, a film-canister lid, and a container of water—to construct a working compass. You will then predict and test how different objects will affect the compass. If your teacher has prepared an orienteering course, you may use your compass to navigate the course.

OBJECTIVES

Construct a compass by using simple materials.

Predict how various objects will affect the compass.

Use the compass to navigate a short orienteering course.

MATERIALS

aluminum can

bar magnets, strong (2)

film-canister lid

index card with directions

iron nail

margarine tub or other small, shallow
   container

marker, permanent

paper clip, large

plastic cup

water

SAFETY

• Use knives and other sharp instruments with extreme care.

Procedure
CONSTRUCT A COMPASS

1. Stroke the paper clip 50 times with one end of a magnet. Stroke in one direction only. Stroking in two directions will not magnetize the paper clip. Be careful not to scrape or puncture yourself with the paper clip.

Name _____ Class _____ Date _____

Constructing and Using a Compass (cont.)

2. Fill the container to the halfway point with water. Gently place a film-canister lid upside down on the surface of the water in the center of the container.

3. Predict what will happen when you set the paper clip on the lid.

4. Test your prediction by gently setting the paper clip on the film-canister lid and observing what happens.

5. Lift the paper clip, and place it back on the lid in a different direction. Repeat this step several times.

6. If all has gone well, you have just made a compass. To determine which end of the paper clip points north, put the south end of the bar magnet about 10 cm from the compass. Use the permanent marker to mark the end of the paper clip that points to the south end of the bar magnet.

7. Carefully remove the paper clip and the canister lid from the water. From this point on, the paper clip will be referred to as the needle of the compass.

8. Use a permanent marker to label all four compass points (N, S, E, and W) on the face of the canister lid.

9. Float the lid in the water. Put the compass needle back on the lid so that the needle points north. You are now ready to use your compass!

Hint: The needle should stay magnetized throughout the activity. However, if it is dropped, it may need to be magnetized again, as in step 1.

TESTING YOUR COMPASS

10. Before you use the compass as a tool, you should discover what might inter-fere with its operation. Predict how each object in the table below might affect the operation of the compass. Record your predictions in a table like the one below.

11. Test your predictions by placing each object 5 cm from the compass (but not in the north or south directions). Record your observations in your data table.

COMPASS RESPONSE DATA

Object	Make a prediction	Conduct an experiment	Make observations
	Will the compass needle move?	Place the object 5 cm from the compass	Did the needle move?
Aluminum can	yes	✔	no
Iron nail	no	✔	yes
Magnet	yes	✔	yes
Plastic cup	no	✔	no

Name _____ Class _____ Date _____

Constructing and Using a Compass (cont.)

USING YOUR COMPASS TO NAVIGATE A COURSE (OPTIONAL)

This part of the Procedure is to be done outdoors in an area prepared by your teacher. In this activity, you will use the compass you have constructed to navigate a course to find objects laid out by your teacher. Your teacher should give you an index card containing directions for you to follow on the course.

12. In many cases, it is impractical to measure distance with a meterstick or a tape measure. Instead, you can measure distance by paces. Your teacher has marked off a 10 m distance. Count your steps as you walk the 10 m. Your steps should be regular and consistent. On the back of the index card, write down the number of steps you walked.

13. Repeat step fifteen twice, and write down the number of steps you walked. In all three cases and in navigating the course, the same person should do the pacing and should take care to keep each step the same length.

14. Calculate your average number of steps in 10 m by adding the total number of steps you took in all three trials and dividing by 3. Write this number down on the back of the index card.

15. Move to the starting point indicated in your directions.

16. Hold the compass, and observe which direction is north.

17. Read the first step of your directions, and determine the direction to walk. Face that direction.

18. Calculate the number of steps required to walk the specified distance. Walk that number of steps.

19. When you reach the destination, you will find an object and an index card containing the name of the object. Record the name of the object in the appropriate place on your index card.

20. Repeat steps 17–19 for each direction on your card until you have listed all five objects.

21. Take your index card to your teacher. If you have successfully completed the course, continue to the Analysis. If you had any difficulty, repeat the course until you are successful.

ANALYSIS

1. Explaining events Describe what you observed when you placed the paper clip on the film-canister lid in step 4. Why do you think this happened?

The canister lid continued to float, and the lid and paper clip rotated to

align the paper clip along the north-south direction. This happened because

the paper clip is magnetized.

Name _____ Class _____ Date _____

Constructing and Using a Compass (cont.)

2. Explaining events How did each of the objects affect the compass in step 11? Explain your results.

The magnet and the nail caused the compass to deflect away from north; the aluminum can and plastic cup had no effect on the compass. The iron nail and magnet have ferromagnetic properties, so the paper clip was attracted to these items. Aluminum and plastic do not have ferromagnetic properties.

CONCLUSIONS

1. Evaluating results Was your prediction in step 3 of the Procedure correct? Why or why not?

Answers may vary. If students predicted that the film-canister lid would continue to float and that the lid and paper clip would rotate, then their prediction was correct.

2. Making predictions What would have happened if you had placed the paper clip on the film-canister lid before you stroked the paper clip with a magnet?

The lid and paper clip would not have turned to point north and south.

EXTENSIONS

1. Building models Imagine that you are lost in the woods without a compass. You have a map of the area, but you do not know which direction is north. It is only a couple of hours before sunset, and you need to get back to your camp. You have with you several topographic maps held together with a paper clip, a bottle of water, a camera and several canisters of film, a plastic storage container of trail mix, and, luckily, a magnet on your key chain. Describe how you could use these materials to build a compass.

I'd eat all of the trail mix and then fill the plastic storage container with water. I'd stroke the paper clip many times in one direction with the magnet and place a film-canister lid upside down on the surface of the water in the container. Finally, I'd set the paper clip on the canister lid. I can use the position of the sun (more west than east) to tell which end of the paper clip points north and which end points south.

Teacher's Notes and Answers

Constructing a Radio Receiver

TIME REQUIRED
2 lab periods

SKILLS ACQUIRED
Constructing models
Experimenting
Inferring
Predicting

THE SCIENTIFIC METHOD

- **Make Observations:** In step 19 of the Procedure, students listen for radio signals and observe how adjusting their receiver affects the operation of the receiver.

- **Analyze the Results:** In the Analysis, students describe the design and explain the function of different parts of the receiver.

- **Draw Conclusions:** In the Conclusions, students compare their simple receivers to a fully functional radio.

- **Communicate the Results:** In the Analysis and Conclusions, students communicate results by providing written answers to the questions.

Teacher's Notes
MATERIALS

Diodes are relatively inexpensive and may be available through your school, a science supply house, or a local electronics store.

Many things may be used as an antenna: a long wire, the antenna from an old stereo, or even the rim and spokes of a bicycle wheel. Inexpensive antennas can also be purchased at electronics stores.

Use large-gauge copper wire (with a small diameter) for the 2 m insulated wires.

The ground wires must be long enough to attach to ground points available in your lab, such as plumbing fixtures.

SAFETY CAUTIONS

Ground wires need to be attached to available ground points in the lab, such as plumbing fixtures.

Students should be instructed to clean earphones between uses, especially if you do not have enough earphones for each student.

DISPOSAL INFORMATION

Instruct students to dissemble their receivers when they are finished. Cardboard tubes, foil, and paper may be thrown into the trash. Tell students where to store the antennas, diodes, earphones, wires, and other materials.

TIPS AND TRICKS

If you do this experiment before students have studied electricity and magnetism, students may not be familiar with the electronic devices used to build the receiver. However, understanding how these devices work beforehand is not necessary. In addition to teaching students about communication technology, this experiment can introduce students to electromagnetism and electronic components, which they will study in greater depth later in the course.

Because the assembly of the receiver has many steps, this activity will probably take two lab periods. Remind students that this lab is long, and encourage them to be patient while making and troubleshooting their receiver. If time or materials are limited, you may choose to do this lab as a special class project or as a demonstration instead.

Remind students that this radio receiver is not very sophisticated and that they probably will not be able to use the receiver to hear music or voice broadcasts. Let students know that even hearing static demonstrates that their receiver is working.

Name _____ Class _____ Date _____

Skills Practice Lab

Constructing a Radio Receiver

Introduction

You have probably listened to radios many times in your life. Modern radios are complicated electronic devices. However, radios do not have to be so complicated. The basic parts of any radio include a diode, a capacitor, an antenna, a ground wire, and an earphone (or a speaker and amplifier in a large radio). In this experiment, you will examine each of these components one at a time as you construct a working radio receiver.

OBJECTIVES

Construct a radio receiver from simple materials.

Use the receiver to detect radio signals.

Explain how different parts of the receiver contribute to the operation of the receiver.

Compare the receiver to a fully functioning radio.

MATERIALS

aluminum foil
antenna
cardboard, 20 cm × 30 cm
cardboard tubes (2)
connecting wires, 30 cm long (7)
diode
earphone

ground wire
insulated wire, 2 m long
paper clips (3)
scissors
sheet of paper
tape

SAFETY

- Secure loose clothing and remove dangling jewelry. Don't wear open-toed shoes or sandals in the lab.

- Do not place electrical cords in walking areas or let cords hang over a table edge in a way that could cause equipment to fall if the cord is accidentally pulled.

- Use knives and other sharp instruments with extreme care.

- Never cut objects while holding them in your hands. Place objects on a suitable work surface for cutting.

Name _____ Class _____ Date _____

Constructing a Radio Receiver (cont.)

Procedure
MAKING AN INDUCTOR AND A CAPACITOR

1. Examine the diode. An electric current in a diode is in only one direction. On a sheet of paper or in your lab notebook, describe the appearance of the diode.

2. An inductor controls the amount of electric current by changing the amount of resistance in a wire. Make an inductor by wrapping a 2 m piece of insulated wire around a cardboard tube approximately 100 times. Leave a few centi-meters of the cardboard exposed on each end. Wind the wire so that each turn of the coil touches but does not overlap the previous turn. Leave about 25 cm of wire loose on each end of the coil. The turns of the coil should end up in a neat and orderly row that extends across most of the length of the tube, as shown in **Figure 1.** Use tape to secure the coil to the tube. Set your inductor aside for now.

FIGURE 1

3. A capacitor stores electrical energy when a current is applied. A variable capacitor is a capacitor in which the amount of stored energy can change. You will make a variable capacitor in steps 4–6.

4. Cut a piece of aluminum foil to go around half the length of another card-board tube. Keep the foil as wrinkle free as possible as you wrap it around the tube. Tape the foil to itself to keep it wrapped around the tube, and then tape the top and bottom of the foil to the tube to keep the foil from sliding.

5. Use the sheet of paper and tape to make a sliding cover over the foil on the tube. The paper should completely cover the foil on the tube with about 1 cm extra. Wrap the paper so that it fits snugly around the foil, but leave it loose enough that it can slide up and down over the foil.

6. Cut another piece of aluminum foil to wrap completely around the paper. Leave approximately 1 cm of paper showing beyond each end of the foil. Keep the foil as wrinkle free as possible as you wrap it around the tube. Do not make the foil so tight that the paper can no longer slide over the bottom layer of foil. Tape the foil to itself to keep it wrapped around the paper, and then tape the top and bottom of the foil to the paper.

7. Stand the tube on end so that the inner layer of foil is on the bottom half. Slide the paper and foil sleeve up so that it covers the top half of the tube.

Name _____ Class _____ Date _____

Constructing a Radio Receiver (cont.)

This tube with the two layers of foil can now serve as a variable capacitor. The amount of stored energy depends on how much the two layers of foil overlap. Therefore, the capacitor stores more energy the farther down the paper and foil sleeve is on the tube. The completed capacitor tube is shown as part of **Figure 2**.

ASSEMBLING THE RECEIVER

8. Use tape to attach a 30 cm connecting wire to the inner foil layer at the bottom of the variable capacitor tube. Also use tape to attach another 30 cm connecting wire to the foil at the bottom of the foil and paper sleeve. Make sure that the metal ends of each wire make good contact with the foil. **Figure 2** shows these wires attached to the capacitor.

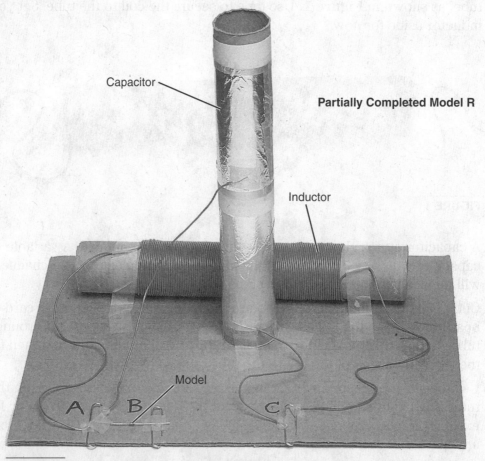

FIGURE 2

9. Hook three paper clips on one edge of a 20 cm × 30 cm piece of cardboard, as shown in **Figure 2**. Label one paper clip "A," another one "B," and the third one "C."

10. Lay the inductor (the tube that you made earlier with the coiled wire) on the cardboard, and tape it to the cardboard, as shown in **Figure 2**.

Name _____ Class _____ Date _____

Constructing a Radio Receiver (cont.)

11. Place the capacitor, still upright in the same orientation, next to the inductor (but not touching it), and tape the bottom of the tube to the cardboard. The tape may cover part of the inner layer of foil, but be careful not to tape the paper and foil sleeve—the sleeve must be free to slide.

12. Use tape to connect the diode to paper clips A and B. The cathode should be closest to paper clip A. (The cathode side of the diode is the side with the dark band; if you are unsure about which direction to place the diode, ask your teacher.) Make sure that all connections have good metal-to-metal contact.

13. Connect the wire from one end of the inductor to paper clip A and the wire from the other end to paper clip C. Use tape to hold the wires in place, and make sure that there is good metal-to-metal contact. You may need to strip some of the insulation off the wire near the tip to get good contact. If so, ask your teacher to help you.

14. Connect the wire from the sliding part of the capacitor to paper clip A. Connect the other wire from the capacitor (the one from the inner foil layer) to paper clip C.

15. Tape a connecting wire to your antenna, and then connect this wire to paper clip A.

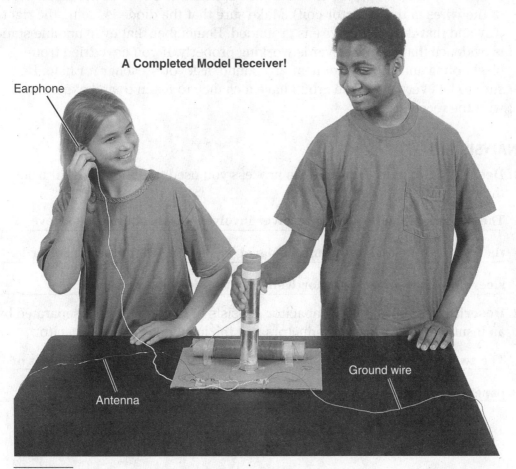

A Completed Model Receiver!

Earphone

Ground wire

Antenna

FIGURE 3

Name _____ Class _____ Date _____

Constructing a Radio Receiver (cont.)

16. Use tape to connect one end of the ground wire to paper clip C. The other end of the wire should be connected to a "ground." This "ground" may be a plumbing fixture or some other point specified by your teacher.

17. Connect one wire from the earphone to paper clip B and the other wire to paper clip C, as shown in **Figure 3.**

USING YOUR RECEIVER

18. Your radio receiver is now complete. It should look similar to the receiver in **Figure 3.** The antenna will pick up radio waves in the air. These waves may be transmitted by a radio station, or they may be "noise" or "static," which are radio waves coming from particles in the atmosphere or even from space. The earphone will allow you to hear the signals (or the noise) corresponding to the radio waves that the antenna picks up.

19. It may take some work to get an audible signal through your receiver. If you do not hear any sound at all, you may need to troubleshoot by checking all of the parts of your receiver one by one. Make sure that there is good contact between metal connections where there is supposed to be contact, and make sure that there is no contact where there is not supposed to be contact (such as between the two pieces of foil in the capacitor or between the metal cores of the wires in the inductor coil). Make sure that the diode is facing the right way and that the ground wire is grounded. Remember that even audible static is evidence that your receiver is working properly. If you have tried troubleshooting and still cannot hear any sound, ask your teacher for help. Be sure to let everyone in the group have a chance to listen to and experiment with the receiver.

ANALYSIS

1. **Describing events** Describe the process you used to obtain a signal from the receiver.

   The process of operating the receiver involved slowly sliding the sleeve of

   the capacitor up and down until a signal was heard. Then, the capacitor

   sleeve was moved back and forth slightly to fine-tune the signal.

2. **Describing events** Every capacitor consists of two conductors separated by an insulator. List the two conductors and the insulator in your capacitor.

   The two conductors are the sheets of foil, and the insulator is the sheet of

   paper.

Name _____ Class _____ Date _____

Constructing a Radio Receiver (cont.)

3. **Explaining events** The amount of energy stored in a capacitor depends in part on the overlapping area of the two plates. As you moved the paper and foil sleeve downward, did the energy storage of the capacitor increase or did it decrease? Explain.

The energy storage of the capacitor increased as the foil and paper sleeve

was moved downward.

4. **Explaining events** One purpose of an inductor in a radio receiver is to increase electrical resistance. The resistance in a wire increases as the length of the wire increases. Why is the resistance in the coil of wire in an inductor greater than the resistance in a straight wire that spans the same length as the inductor?

The resistance in the coil is greater because the total length of wire is

greater.

5. **Explaining events** What kind of wave are the radio waves received by your antenna? What kind of wave passes the signal from the earphone to your ear?

electromagnetic waves; sound waves

CONCLUSIONS

1. **Making predictions** A diode allows current in only one direction. What would happen to your receiver if you turned your diode around so that it faced the wrong way?

The receiver would stop working.

2. **Drawing conclusions** Are the radio waves picked up by your receiver analog signals or digital signals?

analog signals

Name _____ Class _____ Date _____

Constructing a Radio Receiver (cont.)

3. **Evaluating models** List at least two ways that your receiver is similar to a fully functional radio. List at least two ways that your receiver differs from a fully functional radio.

Accept all reasonable answers. The receiver is similar to a fully functional

radio in that it has some of the same electronic components (a diode, a

capacitor, and an inductor), it is tunable, and it uses an antenna to receive

signals. The receiver differs from a fully functional radio in that it has no

power source, amplifier, or volume control, and it cannot detect FM signals.

_____

EXTENSIONS

1. **Research and communications** Research the design of radio transmitters. Write a paragraph that describes some of the materials that you would need in order to build a transmitter to transmit signals that you could pick up with your receiver.

In addition to requiring many of the same materials used in making a

receiver, a radio transmitter requires a power source.

Teacher's Notes and Answers

Explaining the Motion of Mars

TIME REQUIRED
1 lab period

SKILLS ACQUIRED
Classifying
Communicating
Constructing models
Identifying or recognizing patterns
Measuring
Predicting

THE SCIENTIFIC METHOD

- **Ask a question:** In step 1 of the Procedure, students consider why planets move in a back-and-forth pattern.

- **Form a hypothesis:** In step 1 of the Procedure, students form a hypothesis to explain why planets move in a back-and-forth pattern.

- **Analyze the results:** In the Analysis, students analyze the results of the activity.

- **Draw conclusions:** In the Conclusions, students draw conclusions and make predictions based on the results of the activity.

- **Communicate results:** In the Analysis and Conclusions, students communicate results by providing written answers to the questions.

Teacher's Notes

MATERIALS

Provide each lab group with blue, green, and red colored pencils. Make sure the pencils are sharpened. If you do not have enough drawing compasses, groups may share compasses or you may prepare a template showing the orbits of Earth and Mars and make a photocopy for each group.

SAFETY CAUTIONS

Warn students about the sharp point on a drawing compass. If any students have never used a drawing compass, demonstrate how to draw a circle with the compass. Emphasize that the point should be held still and that the paper should be held to keep the paper from sliding.

Name _____ Class _____ Date _____

Explaining the Motion of Mars

Introduction

If you watch the planet Mars every night for several months, you will notice that it appears to "wander" among the stars. While the stars remain in fixed positions relative to each other, the planets appear to move independently of the stars. Mars first travels to the left, then it goes back to the right a little, and finally it reverses direction and travels again to the left. For this reason, the ancient Greeks called the planets "wanderers."

The ancient Greeks believed that the planets and the sun revolved around Earth and that Earth was the center of the solar system. The Greek mathematician Ptolemy came up with a complicated system of "epicycles" that explained why the planets appeared to wander as they orbited Earth (or so he thought).

Ptolemy's model of the solar system was believed for centuries. In the early 1500s, Nicolaus Copernicus proposed that the sun was the center of the solar system and that all the planets, including Earth, revolved around the sun. Copernicus's new model was convincing and was eventually widely accepted because it explained the wandering motion of the planets much better than the Earth-centered model did.

In this lab, you will make your own model of part of the solar system to find out how Copernicus's model of the solar system explained the zigzag motion of the planets.

OBJECTIVES

Illustrate the apparent motion of Mars as seen from Earth against the background of stars.

Explain why Mars moves in a zigzag pattern.

Predict the apparent motion of Jupiter as seen from Earth.

MATERIALS

colored pencils   metric ruler
drawing compass   white paper

Safety

• Use knives and other sharp instruments with extreme care.

Name _____ Class _____ Date _____

Explaining the Motion of Mars (cont.)

Procedure

ASK A QUESTION

1. Why do the planets appear to move back and forth against the background of stars in Earth's night sky? Propose a hypothesis to explain this motion, and write down your hypothesis.

CONDUCT AN EXPERIMENT

2. Mark a point on the center of a piece of paper. Label the point "Sun."

3. Place the point of the compass on the point that represents the sun. Use the compass to draw a circle with a diameter of 9 cm. This circle will represent the orbit of Earth around the sun. (Note: The orbits of the planets are slightly elliptical, but circles will work for this activity.)

4. Using the same center point, draw a circle with a diameter of 12 cm. This circle will represent the orbit of Mars.

5. Using a blue pencil, draw three parallel lines in a diagonal across one end of your paper, as shown in **Figure 1.** These lines will help you plot the path Mars appears to travel in Earth's night sky. Turn your paper so that the diagonal lines are at the top of the page.

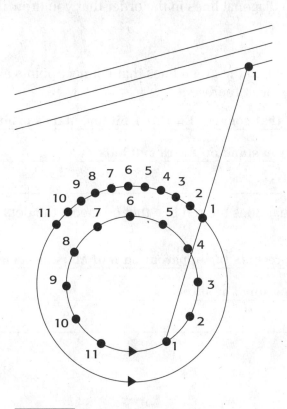

FIGURE 1

6. Place 11 dots on the orbit of Earth and number them 1 through 11 as shown in **Figure 1.** These dots represent Earth's position from month to month.

Name _____ Class _____ Date _____

Explaining the Motion of Mars (cont.)

---

7. Now place 11 dots along the top of the orbit of Mars. Number the dots as shown from month to month. These dots represent the position of Mars in the figure. Notice that Mars travels slower than Earth.

8. Use a green line to connect the first dot on Earth's orbit to the first dot on Mars's orbit, and extend the line until it meets the lowest of the three diagonal blue lines at the top of the paper. Use the ruler as a straight edge to make sure your line is straight. Place a green dot where this green line meets the lowest blue diagonal line, and label the green dot "1."

9. Now connect the second dot on Earth's orbit to the second dot on Mars's orbit, and extend the line until it meets the lowest blue diagonal. Place a green dot where this line meets the lowest blue diagonal line, and label this dot "2."

10. Continue drawing green lines from Earth's orbit to Mars's orbit and up to the blue diagonal lines. Pay attention to the pattern of dots that you are adding to the diagonal lines. When the direction of the dots changes, extend the green line to the middle blue diagonal line, and add the dots to that line instead. If the direction of the dots changes again, extend your green lines to the topmost blue diagonal line, and add the dots to that line instead.

11. When you are finished adding green lines, draw a red line to connect all of the dots on the blue diagonal lines in the order that you drew the dots.

ANALYSIS

1. Classifying What do the green lines that connect points along Earth's orbit and along Mars's orbit represent?

The green lines that connect Earth's orbit and Mars's orbit represent your

line of sight as you stand on Earth and look at Mars.

2. Classifying What does the red line that connects the dots along the diagonal lines represent?

The red line represents the apparent path of Mars as seen from Earth

against the background of stars.

Name _____ Class _____ Date _____

Explaining the Motion of Mars (cont.)

3. Describing events Describe the path of Mars as it would be seen from Earth over the course of the year represented in this activity. Assume Earth is at position 1 in January. In your description, name the months in which Mars would appear to change directions.

Mars would move to the left until the fifth month (May). Then, it would

move to the right until the seventh month (July). After the seventh month, it

would continue to move to the left for the rest of the year.

_____

CONCLUSIONS

1. Drawing conclusions State in your own words why Mars appears to move in a zigzag motion when viewed from Earth against the background of stars.

As the orbit of Earth catches up to the orbit of Mars, Mars appears to move

to the left. When the orbit of Earth passes the orbit of Mars, Mars appears

to reverse its direction and move to the right. After Earth passes Mars, Mars

appears to return to its original direction, moving to the left.

2. Making predictions Do you think the planet Jupiter would appear to move in a zigzag motion when viewed from Earth against the background of stars? Explain why or why not.

Like Mars, Jupiter would appear to move in a zigzag pattern. Jupiter's orbit

around the Sun is outside Earth's orbit, and Jupiter orbits the Sun more

slowly than Earth does. Therefore, Earth will at some point catch up to and

pass Jupiter.

EXTENSIONS

1. Building models Unlike the orbits of Mars and Jupiter, the orbit of Venus around the Sun is inside Earth's orbit. Repeat this experiment, but let the outer circle represent Earth's orbit and the inner circle represent the orbit of Venus. Draw lines from each position of Earth through the corresponding positions of Venus, and extend the lines to the bottom of the page. Would Venus appear to move in a zigzag motion? How would Venus appear to move *relative to the Sun* during the year?

Venus would appear to move in a large zigzag motion. It would oscillate back
and forth around the Sun and would never get very far from the Sun.

Teacher's Notes and Answers

Classifying Rocks

TIME REQUIRED

1 lab period

SKILLS ACQUIRED

Classifying
Collecting data
Experimenting
Communicating
Identifying and Recognizing patterns

THE SCIENTIFIC METHOD

- **Make Observations:** In the Procedure steps 2–4, students make observations of various rock samples.

- **Analyze the Results:** In the Analysis, students examine data and analyze the results of the activity.

- **Draw Conclusions:** In the Conclusions, students draw conclusions and apply their conclusions based on the results of the activity.

- **Communicate Results:** In the Analysis, Conclusions, and Extensions, students communicate results by providing written answers to the questions.

Teacher's Notes

MATERIALS

Before students begin the investigation, number the rocks that they will be identifying. Record the names and numbers of the labeled rocks.

SAFETY CAUTIONS

Remind students to wear goggles and to observe safety requirements when working with the hydrochloric acid.

TIPS AND TRICKS

- Before students begin the investigation, display a few different rock samples. Show at least one igneous rock, one sedimentary rock, and one metamorphic rock. Have students discuss how they would attempt to identify these rocks. Discuss some of the difficulties that students may encounter in trying to identify rocks.

- Review the rock identification table with students. Point out significant characteristics that will help in identification. (For example, rocks whose fragments are cemented together are usually sedimentary rocks.) Tell students to first identify the class to which each rock belongs and to then try to name the rock.

Name _____ Class _____ Date _____

Skills Practice Lab

Classifying Rocks

Introduction

There are many different types of igneous, sedimentary, and metamorphic rocks. Therefore, one must to know important distinguishing features of the rocks in order to classify the rocks. The classification of rocks is generally based on their mode or origin, their mineral composition, and the size and arrangement (or texture) of their minerals. The many types of igneous rocks differ in the minerals that they contain and the sizes of their crystalline mineral grains. Igneous rocks composed of large mineral grains have a coarse-grained texture. Some igneous rocs have small mineral grains that cannot be seen with the unaided eye. These types of rocks have a fine-grained texture.

Many sedimentary rocks are made of fragments of other rocks compressed and cemented together. Some sedimentary rocks have a wide range of sediment sizes, while others may have only one size. Other common features of sedimentary rocks include parallel layers, ripple marks, cross-bedding, and the presence of fossils.

Metamorphic rocks often look similar to igneous rocks, but they may have bands of minerals. Metamorphic rocks with a foliated texture have minerals arranged in bands. Metamorphic rocks that do not have bands of minerals are nonfoliated.

In this investigation, you will use a rock identification table to identify various rock samples.

OBJECTIVES

Observe several properties of rock samples, including color, texture, composition, and whether samples react with acid.

Identify the class and name of each sample of rock by using observations and a rock identification table.

MATERIALS

dilute hydrochloric acid, 10%   rock samples
hand lens   safety goggles
medicine dropper

SAFETY

• If a chemical gets on your skin or clothing or in your eyes, rinse it immediately, and alert your instructor.

• If a chemical is spilled on the floor or lab bench, alert your instructor, but do not clean it up yourself unless your teacher says it is OK to do so.

Name _____ Class _____ Date _____

| *Classifying Rocks (cont.)*

- Wear safety goggles when working around chemicals, acids, bases, flames, or heating devices.

- If any substance gets in your eyes, notify your instructor immediately and flush your eyes with running water for at least 15 minutes.

Procedure

1. List in your data table the numbers of the rock samples that your teacher gave you.

DATA TABLE

Specimen	Description of properties	Rock class	Rock name

2. Using a hand lens, study the rock samples. Look for characteristics such as the shape, size, and arrangement of the mineral crystals. For each sample, list in your table the distinguishing features that you observe.

3. Certain rocks react with acid, which indicates that they are composed of calcite. If a rock contains calcite, it will bubble, releasing carbon dioxide. Using a medicine dropper and 10% dilute hydrochloric acid, test your samples for their reactions. **CAUTION:** *Wear goggles when working with hydrochloric acid.*

4. Refer to the rock identification table. Compare the properties for each rock sample that you listed with the properties listed in the identification table. If you are unable to identify certain rocks, examine these rock samples again.

5. Complete your table by identifying the class of rocks—igneous, sedimentary, or metamorphic—that each rock sample belongs to, and then name the rock.

Name _____ Class _____ Date _____

Classifying Rocks (cont.)

Description	Rock class	Rock name
Coarse-grained; mostly light in color—shades of pink, gray, and white are common	igneous	granite
Coarse-grained; mostly dark in color; much heavier than granite or diorite	igneous	gabbro
Fine-grained; dark in color; often rings like a bell when struck with a hammer	igneous	basalt
Light to dark in color; many holes—spongy appearance; light in weight; may float in water	igneous	pumice
Light to dark in color; glassy luster—sometimes translucent; conchoidal features	igneous	obsidian
Coarse-grained; foliated; layers of different minerals often give a banded appearance	metamorphic	gneiss
Coarse-grained; foliated; quartz abundant; commonly contains garnet; flaky minerals	metamorphic	schist
Fine-grained; foliated; cleaves into thin, flat plates	metamorphic	slate
Coarse-grained; nonfoliated; reacts with acid, effervesces	metamorphic	marble
Fine-grained; soft and porous; normally white or buff in color	sedimentary	chalk
Coarse-grained, over 2 mm; rounded pebbles; some sorting—clay and sand can be seen	sedimentary	conglomerate
Medium-grained, 1/16 mm to 2 mm; mostly quartz fragments—surface feels sandy	sedimentary	sandstone
Microscopic grains; clay composition; smooth surface—hardened mud appearance	sedimentary	shale
Coarse- to medium-grained; well-preserved fossils are common; soft—can be scratched with a knife; occurs in many colors but usually white-gray; reacts with acid	sedimentary	crystalline limestone
Coarse- to fine-grained; cube-shaped crystals; normally colorless; does not react with acid	sedimentary	halite

Name _____ Class _____ Date _____

Classifying Rocks (cont.)

ANALYSIS

1. **Identifying/recognizing patterns** What properties were most useful in identifying each rock sample?

   Answers may vary. Students will probably name texture, structure, and

   mineral composition.

   _____

2. **Classifying** Were any samples difficult to identify? Explain.

   Answers may vary. Students will most likely find fine-grained rocks, such as

   shale, most difficult to identify.

   _____

3. **Classifying** Were any characteristics common to all of the rock samples?

   Answers may vary. All of the rocks contained minerals and were hard.

   _____

   _____

CONCLUSIONS

1. **Drawing conclusions** How can you distinguish between a sedimentary rock and a foliated metamorphic rock when both types of rock have observable layering?

   Answers may vary, but students should realize that many sedimentary rocks

   are made of layers of compressed and cemented sediment particles.

   Metamorphic rocks usually contain bands of minerals.

Name _____ Class _____ Date _____

Classifying Rocks (cont.)

2. **Applying conclusions** Name properties that distinguish between the rocks in the following pairs.

   a. granite and limestone

   Granite has mostly visible mineral crystals; limestone contains calcite,

   which reacts with acid, and individual crystals that are not visible.

   Limestone is softer than granite.

   b. obsidian and sandstone

   Obsidian is made of volcanic glass and has a smooth surface; sandstone

   has a rough, sandy surface.

   c. pumice and slate

   Pumice has many holes and looks spongy; slate splits into thin plates.

   d. conglomerate and gneiss

   Conglomerate has rounded fragments; gneiss has a banded appearance.

EXTENSIONS

1. **Research and communications** Collect a variety of rocks in your area. Use the rock identification table to see how many rocks you can classify. How many rocks are igneous? How many are sedimentary rocks? How many are metamorphic rocks? After you identify the class of each rock, try to name the rock.

   Answers may vary. Students should use their observations to classify and identify the rocks.

Teacher's Notes and Answers

Building a Cup Anemometer

TIME REQUIRED
1 lab period

SKILLS ACQUIRED
Collecting data
Constructing models
Organizing and analyzing data
Predicting

THE SCIENTIFIC METHOD

- **Make Observations:** In step 16 of the Procedure, students measure the rate at which their anemometers rotate.

- **Form a Hypothesis:** In step 15 of the Procedure, students predict the wind speed at the location where they have placed their anemometer.

- **Analyze the Results:** In the Analysis, students calculate wind speeds from measured rotation rates.

- **Draw Conclusions:** In the Conclusions, students compare their wind speed values with their predictions and with the results of other students, and they evaluate the design of their anemometers and suggest improvements.

- **Communicate Results:** In the Conclusions, students communicate results by providing written answers to the questions.

Teacher's Notes

MATERIALS

The straws you provide should be straight straws, not flexible straws. The pencils you provide should be sharp and should have unused erasers. Provide each group with enough modeling clay to make a stable base. If you do not have enough stopwatches to go around, students may use any watch that has a second hand—the time measurements do not need to be very precise.

The second part of the Procedure has the students use their anemometers to measure wind speed. If there is no wind outside, if the wind is very strong (more than 10–15 mi/h), or if there is no safe place for students to set up their anemometers outside, you may set up a fan in the lab. An interesting variation on the lab in this case is to see how much the wind speed changes as distance from the fan increases.

SAFETY CAUTIONS

If students will be measuring wind speeds outside, follow your school's guidelines for outdoor activities.

DISPOSAL INFORMATION

If students are going to do the Extension exercise, provide a safe place for them to keep their anemometers at school. Otherwise, students may dispose of their anemometers or take them home.

Name _____ Class _____ Date _____

Skills Practice Lab

Building a Cup Anemometer

Introduction

An anemometer is a device that measures wind speed. Anemometers are used in weather stations and also on airplanes and ships to measure speed relative to the air. Anemometers have been around for a long time. Drawings by Leonardo da Vinci show his design of a deflection anemometer, in which the wind lifts a plate at an angle corresponding to the wind speed. Many modern anemometers use cups that catch the wind, which causes an axis to rotate at a speed corresponding to the wind speed. This type of anemometer was first designed in 1850 by Dr. Thomas Robinson. In this lab, you will build a simple cup anemometer, and you will use it to measure approximate wind speed.

OBJECTIVES

Construct a cup-style anemometer by using simple materials.

Use an anemometer to measure wind speed.

Compare wind speed measurements with predicted values and with values obtained by others.

Evaluate the design of the anemometer, and suggest improvements.

MATERIALS

hole punch	pencil, sharp with an eraser
marker, colored	plastic straws, straight (2)
masking tape	scissors
metric ruler	stapler
modeling clay	stopwatch
paper cups, small (5)	straight pin

SAFETY

• Secure loose clothing and remove dangling jewelry. Don't wear open-toed shoes or sandals in the lab.

• Use knives and other sharp instruments with extreme care.

• Never cut objects while holding them in your hands. Place objects on a suitable work surface for cutting.

Name _____ Class _____ Date _____

Building a Cup Anemometer (cont.)

Procedure

BUILDING THE ANEMOMETER

1. Cut off the rolled edges of four of the five paper cups. This will make the cups lighter so that they can spin more easily.

2. Make four evenly spaced markings around the outside of the cup that still has its rolled edge. The marks should be 1 cm below the rim of the cup and should go all the way around the cup so that each mark has another mark directly opposite it on the other side of the cup.

3. Use the hole punch to punch a hole at each of the marks on the cup. Use the sharp pencil to carefully punch a hole in the center of the bottom of the cup.

4. Push a straw through two opposite holes in the sides of the cup. Push another straw through the other two opposite holes. The two straws should cross in the center of the cup.

5. Make a mark 3 cm from the bottom on each of the other four cups. At each mark, punch a hole in the cup by using the hole punch.

6. Use the colored marker to color the outside of one of the four cups. The color will allow you to pick out this cup when the cups are spinning.

7. Slide a cup onto one of the straws by pushing the end of the straw through the hole in the cup. Turn the cup so that the bottom of the cup faces to the right when the center cup is facing up.

8. Fold the end of the straw, and staple it to the inside of the cup directly across from the hole (the straw should stretch across the cup).

9. Repeat steps 7–8 for the three remaining cups.

10. Push the pin down through the two straws where they intersect in the middle of the center cup. Use caution because the pin is sharp. Push the eraser end of the pencil up through the bottom of the center cup. Push the pin as far as it will go into the eraser end of the pencil. Be careful not to push too hard—you do not want the pin to bend.

11. Push the sharpened end of the pencil into a solid base of modeling clay. The device should be able to stand upright even when a moderate wind is blowing. Your anemometer should now appear similar to the one shown in **Figure 1.**

12. Blow into the cups so that they spin. Adjust the pin if necessary so that the cups can spin freely without wobbling. Adjust the base if necessary to add stability.

FIGURE 1

Name _____ Class _____ Date _____

Building a Cup Anemometer (cont.)

13. Mark a point on the base with masking tape. Label the tape "Starting point."

USING YOUR ANEMOMETER

14. Find a suitable area outside to place the anemometer on a level surface away from objects that would obstruct the wind, such as buildings or trees. If there is no wind at all or no appropriate place outside, your teacher may set up a fan in the lab.

15. Make a prediction about what the wind speed is at the location where you have placed your anemometer.

16. Hold the colored cup over the starting point in one hand while you or your partner holds the stopwatch. Release the colored cup, and start the stopwatch. As the cups spin, count the number of times the colored cup crosses the starting point in 10 s.

ANALYSIS

1. Describing events How many times did the colored cup cross the starting point in 10 s?

Answers may vary depending on the data.

_____

_____

2. Organizing data Divide your answer in question 1 by 10 to get the average number of revolutions in 1 s.

Answers may vary depending on the data.

_____

_____

3. Organizing data Measure the diameter of your anemometer (the distance from the outer edge of one cup to the outer edge of the opposite cup). Multiply this number by π (≈ 3.14) to get the circumference of your anemometer. Convert the circumference to meters.

Answers may vary depending on the data.

_____

_____

_____

Name _____ Class _____ Date _____

Building a Cup Anemometer (cont.)

4. **Organizing data** Multiply the circumference of the anemometer in meters by the number of revolutions per second that you calculated to get the wind speed in meters per second.

Answers may vary depending on the data.

5. **Organizing data** Convert the wind speed in meters per second to miles per hour. There are 1603 meters in a mile and 3600 seconds in an hour.

Answers may vary depending on the data.

CONCLUSIONS

1. **Evaluating results** Does the value you got for the wind speed seem reasonable? Is it close to what you predicted?

Answers may vary. Students should compare their result to the wind speed

that they predicted in step 15 of the Procedure.

2. **Evaluating results** Compare the wind speed you measured with the wind speeds measured by your classmates. If there are differences, identify at least two potential causes of the differences.

Possible differences include different amounts of friction in the anemom-

eters, errors in measuring the diameter of the anemometer or the number of

rotations in 10 s, and actual differences in wind speed at different times or

locations.

Name _____ Class _____ Date _____

Building a Cup Anemometer (cont.)

3. **Evaluating models** How could you modify your anemometer to make it more accurate or more reliable?

Possible improvements include using stiffer materials, such as wooden

dowels instead of straws and plastic, aluminum cups instead of paper, or a

more stable material for the base; streamlining the closed ends of the cups

so that they have less wind resistance on the leading edge; and calibrating

the anemometer by using a commercial anemometer.

EXTENSIONS

1. **Designing experiments** Use the anemometer to measure wind speed once a day for several days or once a week for several weeks. Check your values against the daily average values given by a local weather station. Are your values consistently higher or lower than the values given by the station? How could you explain this difference? Make a graph of the wind speeds versus time. Do you notice any patterns in the wind speed over time? Try to correlate the wind speeds with other weather events, such as the passage of warm fronts and cold fronts. Write a report that presents your data and summarizes the results.

The Extension exercise encourages students to pay attention to changes in the weather over an extended period of time. Help students find a reliable source of local weather information, such as the National Weather Service Web site. If students do this exercise, provide them with a safe place to keep their anemometers at school and help them choose a safe place where they can take measurements on a regular basis. You may also provide students with additional simple equipment for monitoring the weather, such as rain gauges or thermometers.

Teacher's Notes and Answers

Making Your Own Recycled Paper

TIME REQUIRED
1 lab period

SKILLS ACQUIRED
Constructing models
Experimenting

THE SCIENTIFIC METHOD

- **Make Observations:** During the Procedure, students observe how the materials appear and behave as they use the materials to make paper.

- **Analyze the Results:** In the Analysis, students analyze the changes to the fibers throughout the papermaking process.

- **Draw Conclusions:** In the Conclusions, students evaluate this method of making paper and consider how recycling paper helps the environment.

- **Communicate Results:** In the Analysis and Conclusions, students communicate results by providing written answers to the questions.

Teacher's Notes
MATERIALS
Additional materials for setup

blender	foil
clothesline	glitter
clothespins	flowers and leaves, small
colored thread	other decorations

Set up a "slurry station" in a central location. The station should have a blender and should be near a sink. During the lab, students will bring their paper scraps to the station. You should operate the blender to make the slurry for students. For each group, put 1 L of water in the blender and then add half the contents of the group's cup of paper. Blend until the mixture is evenly blended and has a texture like oatmeal. Pour this slurry into the group's pan, and then add another 1 L of water and the remaining contents of the cup. Blend the mixture, and empty it into the pan.

Also, set up a separate "decoration station" where students can get decorations to add to their slurry. Decorations may include pieces of colored thread of various lengths, small flowers and leaves, small pieces of foil, and glitter.

Set up a drying area. Papers can be dried by hanging them from a clothesline with clothespins. Drying time will vary and may take up to several hours.

You can use almost any kind of paper for the scrap paper that you give the students. Lighter colors work best if students intend to write on their paper.

For Extension exercise

In the Extension exercise, students decorate their paper by using natural materials. You may provide feathers to use for writing, and sponges can be cut up to make stamps. Make the pigments beforehand and keep them in closed containers until they are to be used. When boiled in water, the natural ingredients shown in the table below will make the corresponding pigment colors. Students can experiment with pigments by mixing them together or by diluting them with water. Students should add pigments to the paper only after the paper has dried, so this exercise should be done in a separate lab period.

Ingredient	Color
Turmeric	orange
Marigolds	yellow
Onion peels	yellow
Blueberries	blue
Cranberries	red
Raspberries	pink

SAFETY CAUTIONS

None of the materials in this lab are caustic or dangerous. However, some materials, especially the pigments, may stain clothing or skin. Make sure you and your students wear lab aprons and disposable gloves when working with pigments.

Remind students to use extra caution when cutting paper. Because they are cutting the paper into small pieces, their risk of cutting themselves is greater.

DISPOSAL INFORMATION

Provide a special large container for students to dispose of unused slurry. This unused slurry can be put in a compost pile or thrown away with solid waste. Do not pour slurry down the drain. Slurry can clog pipes.

TECHNIQUES TO DEMONSTRATE

You may wish to demonstrate steps 7–10 of the Procedure, in which the embroidery hoop with nylon is dipped into slurry, pulled out and drained, and then flipped over onto a towel and dried. You can use slurry that you have prepared before class, or you can demonstrate earlier steps in the Procedure, including making the slurry.

Name _____ Class _____ Date _____

Skills Practice Lab

Making Your Own Recycled Paper

Introduction

Paper has been made of many different materials throughout history. In the old days, Europeans often used animal skin to make paper. The Egyptians used grass. Paper in Asia is often made from rice. Most paper in the United States today is made from trees, which can take a toll on the environment. Your city or town may have a paper recycling program. However, even commercial recycling of paper uses precious energy resources, and the recycled paper is often supplemented with fresh wood pulp.

One easy and fun way that you can help the environment is to make your own recycled paper from paper scraps. You can even incorporate other fibers or natural dyes into your paper to make the paper more artistic. You can then use your homemade paper to write letters, send greeting cards, or keep a journal. In this activity, you will learn the basics of making your own paper. Once you know how, you can experiment to find unique ways to customize and decorate your paper.

OBJECTIVES

Create recycled paper from paper scraps.

Describe how paper fibers change throughout the papermaking process.

Explain how recycling paper can help the environment.

Evaluate this method of making recycled paper.

MATERIALS

cornstarch, 5 mL
cotton balls
cup, 500 mL
duct tape
embroidery hoop, 20 cm to 26 cm
    in diameter
long-handled spoon
nylon stocking, no runs

roasting pan, at least 10 cm deep
scissors
scrap paper
sponges (3 or 4)
towels (3 or 4)
water

SAFETY

- Secure loose clothing and remove dangling jewelry. Don't wear open-toed shoes or sandals in the lab.

- Always use caution when working with chemicals.

- Never mix chemicals unless specifically directed to do so.

Name _____ Class _____ Date _____

Making Your Own Recycled Paper (cont.)

- Wear an apron or lab coat to protect your clothing when working with chemicals.

- Follow instructions for proper disposal.

- If a spill gets on your clothing, rinse it off immediately with water for at least 5 minutes while notifying your instructor.

- Always wear protective gloves when handling chemicals.

- Do not eat any part of a plant or plant seed used in the lab.

- Wash hands thoroughly after handling any part of a plant.

- Use knives and other sharp instruments with extreme care

- Never cut objects while holding them in your hands. Place objects on a suitable work surface for cutting.

- Never use a double-edged razor in the lab.

Procedure

1. Separate the inner and outer parts of the embroidery hoop. Put the inner hoop into the stocking, being careful not to rip the stocking or cause a run. When the hoop is completely covered, replace the outer hoop and tighten the screw so that it fits snugly. The hoop is now a papermaker.

2. Place four pieces of duct tape along the edges of the hoop so that the center forms a rectangle, as shown in **Figure 1.**

3. Add 5 mL of cornstarch to a 500 mL cup. Adding cornstarch to the paper will prevent ink from bleeding. Shred paper into small pieces (very roughly 1 inch square). Loosely pack the shreds into the cup until the cup is full.

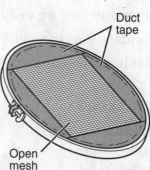

Duct tape

Open mesh

FIGURE 1

4. Your teacher should have set up a central station for blending your paper scraps into a smooth pulp, called *slurry*. At the slurry station, carefully add 1 L of water to the blender. Add half of the contents of the cup to the blender, and securely place the lid on the blender. Your teacher will blend the mixture until it forms into slurry. Carefully pour the slurry into a pan that is at least 10 cm deep. Use the remainder of the cup's contents to repeat this process.

5. Your teacher should have set up another station that contains various items for decorating your paper. At the decoration station, choose various items to incorporate into your paper. Add them to your slurry, and blend the mixture with a spoon. Be careful not to add too much decorative material; such material could cause the paper to be too weak and to tear easily.

6. Separate the fibers of several cotton balls, and add them to the slurry to strengthen the paper. Stir the fibers into the slurry with a spoon until they are evenly distributed.

Name _____ Class _____ Date _____

Making Your Own Recycled Paper (cont.)

7. Use tape to label a towel with your name. Unfold the towel, and place it beside the pan. Hold the papermaker tape-side up. Scoop the slurry into the papermaker, and let the slurry rest for a few minutes on the bottom of the pan. Fibers will settle on top of the papermaker.

8. Without tilting the papermaker, slowly lift it out of the pan. Let the water drain into the pan. When the papermaker stops dripping, carefully flip it onto the towel so that the new paper lies between the papermaker and the towel.

9. Gently press on the nylon with the sponge, and rub the sponge along the back of the mesh to absorb the water. Removing the excess water strengthens your paper and helps it dry quickly. When your sponge is full of water, wring the water into the pan. Repeat this process until you can no longer remove water from the paper.

10. Carefully lift the papermaker off your new paper sheet. Gently lift the towel with the paper on it, and move the towel and the paper to the drying area. (Hint: If the paper starts to rip, press the edges of the tear together with wet fingers.) Congratulations! You have made your own paper!

11. Repeat steps 6–10 for each group member.

12. Discard any unused pulp mixture into a compost pile or a container that your teacher has provided. Do not pour the slurry down the drain. Slurry clogs pipes.

ANALYSIS

1. **Describing events** The paper you have made is composed of dried plant fibers. Fill in the table below to describe what you think happened to the fibers in each of the steps that you followed.

Action	What happened to the fibers?
Added water to the scrap paper	The fibers absorbed water and became mushy.
Blended the scrap paper to make slurry	The blender cut the fibers, which made them shorter.
Lifted the papermaker from the pan	The fibers were caught in the nylon mesh.
Dried the paper	The meshed fibers contracted and became stuck together.

Name _____ Class _____ Date _____

Making Your Own Recycled Paper (cont.)

2. Analyzing results How many pieces of paper did your group originally use to make the scraps that became your slurry? How many pieces of recycled paper did your group make?

Answers may vary.

_____

_____

_____

CONCLUSIONS

1. Drawing conclusions How does recycling paper help the environment?

As more used resources are recycled, there is less demand for new ones.

Therefore, making recycled paper helps conserve materials used to make

paper, such as wood pulp from trees.

_____

2. Evaluating results How could this lab activity be modified to better conserve resources?

Sample answer: by using thinner slurry (which would require less pulp), by

using the leftovers from other groups, and by collecting paper scraps from

local homes or businesses.

_____

EXTENSIONS

1. Designing experiments If you have time and if your teacher has provided the materials, you can use natural materials to decorate your paper once it has dried. Your teacher should provide pigments made from flowers, fruit, vegetables, or spices. Be sure to wear a lab apron and disposable gloves while working with the pigments. You can use sponges to create textured patterns with the pigments on your paper. You can cut the sponges into different shapes to make stamps and then use the stamps to make repeated shapes on the paper. You can also use a feather dipped in pigment to write or draw on the paper. Let the pigments dry before moving or handling the paper.

Laboratory Manual

HOLT, RINEHART AND WINSTON

A Harcourt Education Company

Austin • Orlando • Chicago • New York • Toronto • London • San Diego

Holt Science Spectrum
Laboratory Manual

Cover: Pete Saloutos/Corbis Stock Market

Unless otherwise noted all other illustrations by Holt, Rinehart and Winston.

Page 47 (Fig. A), Thomas Gagliano; 65–67 (Fig. 1-4), Thomas Gagliano; 92 (Fig. 1), Mark Heine.

Printed in the United States of America

ISBN 0-03-067076-4

1 2 3 4 5 6 082 06 05 04 03 02

Contents

Writing a Laboratory Report

In many of the laboratory investigations that you will be doing, you will be asking a question and then performing experiments to find out the answer to your question. Laboratory reports for experimental investigations like these should contain the following parts:

TITLE

This is the name of the laboratory investigation you are doing. If you are performing an investigation from a laboratory manual, the title will be the same as the title of the investigation.

HYPOTHESIS

The hypothesis is what you think will happen during the investigation. It is often posed as an "If...then" statement. When you do your experiment, you will be changing one condition, or variable, and observing and measuring the effect of this change. The condition that you are changing is called the independent variable and should follow the "If..." statement. The effect that you expect to observe is called the dependent variable and should follow the "...then" statement. For example, "if salamanders are reared in acidic water" (the independent variable—salamanders normally live in nearly neutral water and you are changing this to acidic water), "then more salamanders will develop abnormally" (the dependent variable—this is the change that you expect to observe and measure).

MATERIALS

This is a list of all the equipment and other supplies you will need to complete the investigation. If the investigation is taken from a laboratory manual, the materials are generally listed for you.

PROCEDURE

The procedure is a step-by-step explanation of exactly what you did in the investigation. Investigations from laboratory manuals will have the procedure carefully written out for you. The procedure section should include an explanation of why your experiment was a controlled experiment. In a controlled experiment, you will change a single independent variable in an experimental group, and compare the results to those obtained in a control group, in which the independent variable was not changed. For example, if you wish to see if acidic water increases the number of abnormalities among salamanders, your experiment must include a control group of salamanders raised under normal conditions so that you will know what percentage of salamanders reared under normal conditions are abnormal.

DATA

Your data are your observations. They are often recorded in the form of tables, graphs, and drawings.

ANALYSES AND CONCLUSIONS

This part of the investigation explains what you have learned. You should evaluate your hypothesis and explain any errors you made in the investigation. Keep in mind that not all hypotheses will be correct. That is normal. You simply need to explain why things did not work out the way you thought they would. In laboratory manual investigations, there will be questions to guide you in analyzing your data. You should use these questions as a basis for your conclusions.

Safety in the Laboratory

Systematic, careful lab work is an essential part of any science program because lab work is the key to progress in science. In this class, you will practice some of the same fundamental laboratory procedures and techniques that scientists use to pursue new knowledge.

The equipment and apparatus you will use involve various safety hazards, just as they do for working scientists. You must be aware of these hazards. Your teacher will guide you in properly using the equipment and carrying out the experiments, but you must also take responsibility for your part in this process. With the active involvement of you and your teacher, these risks can be minimized so that working in the laboratory can be a safe, enjoyable process of discovery.

Anything can be dangerous if it is misused. Always follow the instructions for the experiment. Pay close attention to the safety notes. Do not do anything differently unless told to do so by your teacher. If you follow the rules stated below, pay attention to your teacher's directions, and follow the cautions listed on chemical labels, equipment, and in the experiments, then you will stay safe.

THESE SAFETY RULES ALWAYS APPLY IN THE LAB

1. **Wear safety goggles, gloves, and a laboratory apron.**
   Wear these safety devices whenever you are in the lab, not just when you are working on an experiment. Even if you aren't working on an experiment, laboratories contain chemicals that you should protect yourself from. Keep the lab apron strings tied.

   If your safety goggles are uncomfortable or cloud up, ask your teacher for help. Try lengthening the strap, washing the goggles with soap and warm water, or using an anti-fog spray.

2. **No contact lenses in the lab.**
   Contact lenses should not be worn during any investigations using chemicals (even if you are wearing goggles). In the event of an accident, chemicals can get behind contact lenses and cause serious damage before the lenses can be removed. If your doctor requires that you wear contact lenses instead of glasses, you should wear eye-cup safety goggles in the lab. Ask your doctor or your teacher how to use this very important and special eye protection.

3. **NEVER work alone in the lab.**
   Work in the lab only while under the supervision of your teacher. Do not leave equipment unattended while it is in operation.

4. **Wear the right clothing for lab work.**
   Necklaces, neckties, dangling jewelry, long hair, and loose clothing can get caught in moving parts or catch on fire. Tuck in neckties or take them off. Do not wear a necklace or other dangling jewelry, including hanging earrings. It might also be a good idea to remove your wristwatch so that it is not damaged by a chemical splash. Wear shoes that will protect your feet from chemical spills and falling objects—no open-toed shoes or sandals, and no shoes with woven leather straps.

5. **Only books and notebooks needed for the experiment should be in the lab.**

Only the lab notebook and perhaps the textbook should be used. Keep other books, backpacks, purses, and similar items in your desk or locker.

6. **Read the entire experiment before entering the lab.**

Memorize the safety precautions. Be familiar with the instructions for the experiment. Only materials and equipment authorized by your teacher should be used. Your teacher will review any applicable safety precautions before the lab. If you are not sure of something, ask your teacher about it.

7. **Always heed safety symbols and cautions listed in the experiments, on handouts, and those posted in the room and given verbally by your teacher.**

They are provided for a reason: YOUR SAFETY.

8. **Read chemical labels.**

Follow the instructions and safety precautions stated on the labels.

9. **Be alert and walk with care in the lab.**

Sometimes you will have to carry chemicals *from* the supply station to your lab station. Avoid bumping into other students and spilling the chemicals. Stay at your lab station at other times. Be aware of others near you or your equipment when you are about to do something. If you are not sure of how to proceed, ask.

10. **Know the proper fire drill procedures and location of fire exits and emergency equipment.**

Make sure you know the procedures to follow in case of a fire or emergency.

11. **Know the location and operation of safety showers and eyewash stations.**

12. **If your clothing catches on fire, do not run; WALK to the safety shower, stand under it, and turn it on.**

Call your teacher while you do this.

13. **If you get a chemical in your eyes, walk immediately to the eyewash station, turn it on, and lower your head so your eyes are in the running water.**

Hold your eyelids open with your thumbs and fingers, and roll your eyeballs around. You have to flush your eyes continuously for at least 15 minutes. Call your teacher while you are doing this.

14. **If you spill a chemical on your skin, wash it off with lukewarm water, and call your teacher.**

If you spill a solid chemical on your clothing brush it off carefully without scattering it on somebody else, and call your teacher. If you get liquid on your clothing, wash it off right away using the sink faucet, and call your teacher. If the spill is on your pants or somewhere else that will not fit under the sink faucet, use the safety shower. Remove the pants or other affected clothing while under the shower, and call your teacher. (It may be temporarily embarrassing to remove pants or other clothing in front of your class, but failing to flush that chemical off your skin could cause permanent damage.)

15. **Report all accidents to the teacher immediately, no matter how minor.**
In addition, if you get a headache, feel sick to your stomach, or feel dizzy, tell your teacher immediately.

16. **Report all spills to your teacher immediately.**
Call your teacher rather than trying to clean a spill yourself. Your teacher will tell you if it is safe for you to clean up the spill; if not, your teacher will know how the spill should be cleaned up safely.

17. **The best way to prevent an accident is to stop it before it happens.**
If you have a close call, tell your teacher so that you and your teacher can find a way to prevent it from happening again. Otherwise, the next time, it could be a harmful accident instead of just a close call.

18. **Student-designed inquiry investigations, such as the Design Your Own Labs in the textbook, must be approved by the teacher before being attempted by the student.**

19. **DO NOT perform unauthorized experiments or use equipment and apparatus in a manner for which they were not intended.**
Use only materials and equipment listed in the activity equipment list or authorized by your teacher. Steps in a procedure should only be performed as described unless your teacher gives you approval to do otherwise.

20. **Food, beverages, chewing gum, and tobacco products are NEVER permitted in the lab.**

21. **For all chemicals, take only what you need.**
However, if you happen to take too much and have some left over, DO NOT put it back into the container. Ask your teacher what to do with any leftover chemicals.

22. **NEVER taste chemicals. Do not touch chemicals or allow them to contact areas of bare skin.**

23. **Use a sparker to light a Bunsen burner.**
Do not use matches. Be sure that all gas valves are turned off and that all hot plates are turned off and unplugged when you leave the lab.

24. **Use extreme caution when working with hot plates or other heating devices.**
Keep your head, hands, hair, and clothing away from the flame or heating area, and turn the devices off when they are not in use. Remember that metal surfaces connected to the heated area will become hot by conduction. Remember that many metal, ceramic, and glass items do not always look hot when they are hot. Allow all items to cool before storing.

25. **Do not use electrical equipment with frayed or twisted wires.**

26. **Be sure your hands are dry before using electrical equipment.**
Before plugging an electrical cord into a socket, be sure the electrical equipment is turned OFF. When you are finished with the device, turn it off. Before you leave the lab, unplug the device, but be sure to turn it off FIRST.

27. **Do not let electrical cords dangle from work stations; dangling cords can cause tripping or electrical shocks.**

    The area under and around electrical equipment should be dry; cords should not lie in puddles of spilled liquid.

28. **Horseplay and fooling around in the lab are very dangerous.**

    Laboratory equipment and apparatus are not toys; never play in the lab or use lab time or equipment for anything other than their intended purpose.

29. **Keep work areas and apparatus clean and neat.**

    Always clean up any clutter made during the course of lab work, put away apparatus in an orderly manner, and report any damaged or missing items.

30. **Always thoroughly wash your hands with soap and water at the conclusion of each investigation.**

31. **Whether or not the lab instructions remind you, all of these rules apply all of the time.**

SAFETY SYMBOLS

The following safety symbols will appear in the laboratory manual to emphasize important additional areas of caution. Learn what they represent so you can take the appropriate precautions. Remember that the safety symbols represent hazards that apply to a specific activity, but the numbered rules given on the previous pages always apply to all work in the laboratory.

Animal Safety

- Always obtain permission before bringing any animal to school.

- Handle animals carefully and with respect.

- Wash your hands thoroughly after handling any animal.

Caustic Substances

- If a chemical gets on your skin or clothing or in your eyes, rinse it immediately, and alert your instructor.

- If a chemical is spilled on the floor or lab bench, alert your instructor, but do not clean it up yourself unless your teacher says it is OK to do so.

Chemical Safety

- Always use caution when working with chemicals.

- Never mix chemicals unless specifically directed to do so.

- Never taste, touch, or smell chemicals unless specifically directed to do so.

- Add acid or base to water; never do the opposite.
- Never return unused chemicals to the original container.
- Never transfer substances by sucking on a pipette or straw; use a suction bulb.
- Follow instructions for proper disposal.

Clothing Protection

- Secure loose clothing and remove dangling jewelry. Donít wear open-toed shoes or sandals in the lab.
- Wear an apron or lab coat to protect your clothing when working with chemicals.
- If a spills gets on your clothing, rise it off immediately with water for at least 5 minutes while notifying your instructor.

Electrical Safety

- Do not place electrical cords in walking areas or let cords hang over a table edge in a way that could cause equipment to fall if the cord is accidentally pulled.
- Do not use equipment with damaged cords.
- Be sure that equipment is in the "off" position before plugging it in.
- Never use an electrical appliance around water.
- Be sure to turn off and unplug electrical equipment when finished

Eye Protection

- Wear safety goggles when working around chemicals, acids, bases, flames or heating devices. Contents under pressure may become projectiles and cause serious injury.
- Never look directly at the sun through any optical device or use direct sunlight to illuminate a microscope.
- Avoid wearing contact lenses in the lab.
- If any substance gets in your eyes, notify your instructor immediately and flush your eyes with running water for at least 15 minutes.

Explosion Safety

- Use flammable liquids only in small amounts.

- When working with flammable liquids, be sure that no one else in the lab is using a lit Bunsen burner or plans to use one. Make sure there are no other heat sources present.

Glassware Safety

- Check the condition of glassware before and after using it. Inform your teacher of any broken, chipped, or cracked glassware because it should not be used.

- Do not pick up broken glass with your bare hands. Place broken glass in a specially designated disposal container.

Hand Safety

- In order to avoid burns, wear heat-resistant gloves whenever instructed to do so.

- Always wear protective gloves when handling chemicals.

- If you are unsure of whether an object is hot, do not touch it.

- Use tongs when heating test tubes. Never hold a test tube in your hand to heat it.

Heating Safety

- Avoid wearing hair spray or hair gel on lab days.

- Whenever possible, use an electric hot plate as heat source instead of an open flame.

- When heating materials in a test tube, always angle the test tube away from yourself and others.

- Glass containers used for heating should be made of heat-resistant glass.

- Know your school's fire-evacuation routes.

Plant Safety

- Do not eat any part of a plant or plant seed used in the lab.
- Wash hands thoroughly after handling any part of a plant.
- When in nature, do not pick any wild plants unless your teacher instructs you to do so.

Sharp Object

- Use knives and other sharp instruments with extreme care.
- Never cut objects while holding them in your hands. Place objects on a suitable work surface for cutting.
- Never use a double-edged razor in the lab.

Waste Disposal

- Clean and decontaminate all work surfaces and personal protective equipment as directed by your instructor.
- Dispose of all sharps (broken glass and other contaminated sharp objects) and other contaminated materials (biological and chemical) in special containers as directed by your instructor.

Name _____ Class _____ Date _____

Skills Practice Lab

Comparing the Densities of Pennies

Introduction

All pennies are the same, right? All pennies are about the same size and color—and all are worth one cent—but not all pennies have exactly the same physical properties. Individual pennies may differ from one another based on how much they have corroded or how much dirt they have accumulated. A more significant difference may be in the composition of the metals that make up the pennies. Occasionally, the U.S. Department of the Treasury changes the composition of pennies, which changes the overall mass and density of each coin. In this experiment, you will determine the average density of two group of pennies made at different times. Then, you will compare the average densities of the pennies in the two groups.

OBJECTIVES

Construct the average densities of pennies by using mass and volume measurements.

Compare the average densities of pennies made before 1982 and after 1982.

Infer the change made in the composition of pennies in 1982.

MATERIALS

balance   paper towels
graduated cylinder, 100 mL   pennies (10)
paper   water

Safety

- Secure loose clothing and remove dangling jewelry. Don't wear open-toed shoes or sandals in the lab.

- Check the condition of glassware before and after using it. Inform your teacher of any broken, chipped, or cracked glassware because it should not be used.

- Do not pick up broken glass with your bare hands. Place broken glass in a specially designated disposal container.

Procedure

1. Sort your pennies into two piles. One pile should contain pennies made before 1982. The other pile should contain pennies made after 1982. If you have a penny made in 1982, ask your teacher to replace it with a different penny. Record the number of coins in each pile in a data table like the one shown on the next page.

Comparing the Densities of Pennies (cont.)

2. Use the balance to measure the combined mass of the pennies in the pre-1982 group. Record the mass in grams in your data table. Then, use the balance to measure the combined mass of the pennies in the post-1982 group. Record the mass in grams in your data table.

3. Fill a graduated cylinder about halfway with water. Note that the top surface of the water curves upward toward the edges of the cylinder. The curved surface of the water is called a *meniscus*. You should always use the bottom of the meniscus when measuring volume in a graduated cylinder, and you should always put your eye at the same level as the meniscus when taking your volume reading.

4. Measure the volume of water in the graduated cylinder by using the bottom of the meniscus. Record the volume in milliliters in your data table.

5. Now add the group of pre-1982 pennies to the water in the graduated cylinder. Measure the new volume of the water and pennies combined. Record the volume in milliliters in your data table.

6. Carefully remove the pennies from the graduated cylinder, and pour the water into a sink. Use a paper towel to dry the pennies.

7. Repeat steps 3–6 for the group of post-1982 pennies.

8. When you are finished, place the graduated cylinder on a rack to dry and return the pennies to your teacher.

DATA TABLE

Group	# of coins	Volume of mass (g)	Volume of water (mL)	Average water and coins (mL)	Coins (mL)	Density of coins (g/mL)
Pre-1982						
Post-1982						

ANALYSIS

1. Organizing data Calculate the volume of the pre-1982 pennies by subtracting the volume of the water from the combined volume of the water and coins. Record this volume in your data table. Repeat the calculation for the post-1982 pennies, and record the result.

_____

_____

_____

_____

2. Organizing data Calculate the average density of the pre-1982 pennies by dividing the mass of the coins by the volume of the coins and then dividing the result by the number of coins in the group. Record the average density in your data table. Repeat the calculation for the post-1982 pennies, and record the result.

_____

_____

_____

_____

3. Analyzing results Which group of pennies has a greater average density?

_____

_____

_____

_____

CONCLUSIONS

1. Drawing conclusions Why do you think that pennies made before 1982 have a different density than pennies made after 1982 do?

_____

_____

_____

_____

2. Applying conclusions Pennies made before 1982 and pennies made after 1982 contain a combination of copper and zinc. The density of copper is greater than the density of zinc. Is the ratio of copper to zinc in the pennies made after 1982 higher or lower than the ratio of copper to zinc in the pennies made before 1982?

_____

_____

_____

_____

Name _____ Class _____ Date _____

EXTENSIONS

1. Research and communications Research changes in the composition of pennies made by the U.S. Department of the Treasury in the past century. Write a paragraph describing the changes and explaining why you think the changes were made.

Name _____ Class _____ Date _____

Measuring Density with a Hydrometer

Introduction

A hydrometer is a device used for measuring the density of a liquid. Hydrometers have a wide variety of uses. For example, beverage makers use a hydrometer to test their product at different stages during production. The density of the beverage serves to indicate whether certain chemical processes have taken place in the production process. An ecologist can use a hydrometer to test the salinity of water. This test tells the ecologist whether the water is a suitable habitat for different kinds of aquatic organisms. In this experiment, you will construct a simple hydrometer. You will not use your hydrometer to measure exact values. Instead, you will use the hydrometer to compare the densities of fresh water and salt water and to compare the densities of regular cola and diet cola.

OBJECTIVES

Construct reaction time of a falling object by using falling-distance data.

Compare reaction times measured by using a stopwatch with reaction times calculated from falling-distance data.

Explain the methods used to determine reaction time in the experiment.

MATERIALS

balance	paper towels
beakers, large (4)	plastic drinking straws
cola, both regular and diet	salt
colored permanent markers	scissors
masking tape	stirring rod
metal BBs (6)	water
modeling clay	

SAFETY

- Secure loose clothing and remove dangling jewelry. Don't wear open-toed shoes or sandals in the lab.

- Do not pick up broken glass with your bare hands. Place broken glass in a specially designated disposal container.

- Check the condition of glassware before and after using it. Inform your teacher of any broken, chipped, or cracked glassware because it should not be used.

- Use knives and other sharp instruments with extreme care. Never cut objects while holding them in your hands. Place objects on a suitable work surface for cutting.

Procedure

1. Fill a beaker about half full with water. Label this beaker "Fresh water" by using the marker and a piece of masking tape.

2. Fill another beaker about half full with water. Pour about 4 Tbsp of salt into the beaker, and stir with a stirring rod. Label this beaker "Salt water."

3. Fill another beaker about half full with regular cola, and label the beaker "Regular cola." Fill a fourth beaker with diet cola, and label the beaker "Diet cola." Caution: Do not drink any of the cola or eat or drink anything else in the lab.

4. Cut a plastic drinking straw in half. Seal one end of the straw with modeling clay. Hold the straw vertically so that the open end of the straw points up. Drop two BBs into the open end of the straw. You now have a simple hydrometer.

5. Place the hydrometer, clay end first, into the beaker of fresh water. Adjust the hydrometer by adding or removing BBs until the water level is about halfway up the body of the hydrometer.

6. Use a marker to mark the water level on the hydrometer when the hydrometer is in fresh water. Pin the hydrometer against the side of the beaker, and hold the hydrometer in place with your fingers).

7. Predict whether the salt water is denser or less dense than the fresh water? Will this difference in density cause the hydrometer to sink more or less into the salt water than into the fresh water? Discuss your prediction with the other members in your lab group.

8. Remove the hydrometer from the fresh water. Place the hydrometer, clay end down, into the beaker of salt water. Mark the water level on the hydrometer by using a marker of a different color than the marker you used for the fresh water line. (Tip: You can mark your beaker labels with the colored markers so that you can remember which color corresponds to which liquid.

9. Predict whether regular cola or diet cola is denser. In which liquid will the hydrometer sink the least?

10. Repeat step 8 for the beakers of regular cola and diet cola. Use a different colored marker to mark the water level in each beaker.

ANALYSIS

1. **Organizing data** List in order from bottom to top the colors of the lines on the hydrometer. Then repeat the list, substituting the names of the corresponding liquids for the colors.

_____

_____

_____

Measuring Density with a Hydrometer (cont.)

2. Describing events Did the hydrometer sink lower in the fresh water or in the salt water? How does this result compare with the prediction you made in step 7 of the Procedure?

_____

_____

_____

_____

3. Describing events Did the hydrometer sink lower in the regular cola or in the diet cola? How does this result compare with the prediction you made in step 9 of the Procedure?

_____

_____

_____

_____

4. Explaining events Buoyancy is the force with which a liquid pushes upward on a less dense substance. Buoyancy increases as the density of the liquid increases. If a hydrometer is moved from one liquid to a second liquid that has greater density than the first liquid, would the hydrometer sink into the second liquid more or less than it sank into the first liquid?

_____

_____

_____

_____

CONCLUSIONS

1. Drawing conclusions Which is denser: fresh water or salt water? Explain why. Use the concepts of mass and volume in your explanation. Compare your answer with your prediction in step 7 of the Procedure.

_____

_____

_____

_____

Measuring Density with a Hydrometer (cont.)

2. **Drawing conclusions** Which is denser: regular cola or diet cola? Form a hypothesis to explain your answer. Use the concepts of mass and volume in your explanation. Compare your answer with your prediction in step 9 of the Procedure.

_____

_____

_____

_____

_____

3. **Applying conclusions** If you placed a can of diet cola and a can of regular cola in a bucket of fresh water, what would happen to the cans? Explain your answer.

_____

_____

_____

_____

EXTENSIONS

1. **Designing experiments** Design an experiment to compare the density of fresh cola with the density of flat cola. Predict which cola will have a greater density.

Skills Practice Lab

Drawing Atomic Models

Introduction

Look at the many different things in your classroom or lab: desks, chairs, windows, laboratory equipment, students, shoes, and notebooks. If all of these things are made from atoms and all atoms are made of only a few kinds of particles, what accounts for the variety of things that you see?

Atoms of different elements have different numbers of protons in their nuclei. In atoms that have a neutral charge, the number of electrons around the nucleus equals the number of protons in the nucleus. As it turns out, the number of electrons in an atom is one of the most important things in determining the chemical properties of the atom.

Elements are arranged in the periodic table according to the number of protons in each atom, which also corresponds to the number of electrons in the atom. The periodic table is called "periodic" because as the number of protons increases, certain chemical properties appear over and over again—periodically. In this activity, you will draw models of atoms and will place the atoms in the proper place on a periodic table.

OBJECTIVES

Draw models of atoms that show the numbers of protons and neutrons and electrons in proper energy levels.

Locate the proper position of atoms on a periodic table.

Infer the relationship of the number of energy levels and number of valence electrons in an atom to the group and period of the atom on a periodic table.

MATERIALS

paper   periodic table of the elements
pencil

Procedure

DRAWING MODELS OF ATOMS

1. **Figure 1** shows a model of an atom. This type of model is sometimes called a *Bohr model* because it was first used by the physicist Niels Bohr. The model shows a nucleus in the center that has three protons (p^+) and four neutrons (n). Surrounding the nucleus are three electrons (e^-). Notice that the number of electrons equals the number of protons in the nucleus.

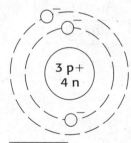

FIGURE 1

2. The atomic number of an atom equals the number of protons in the atom's nucleus. The sum of the number of protons and the number of neutrons in the nucleus is the mass number of an atom. In a data table like **Data Table 1** shown below, fill in the number of protons, the number of neutrons, the atomic number, and the mass number for the atom in **Figure 1** (Atom A).

3. Notice that **Figure 1** shows the first two electrons in Atom A in the first energy level, on the circle closest to the nucleus. The third electron is in the second energy level, on a circle farther from the nucleus. The first energy level of any atom can hold only up to two electrons. The second and third energy levels can each hold up to eight electrons. In a data table like **Data Table 2** shown below, fill in the total number of electrons, the number of energy levels, and the number of electrons in the highest energy level (the circle farthest from the nucleus) for Atom A.

4. On a separate piece of paper, draw a model of an atom that has 11 protons and 12 neutrons. Remember that the first energy level for electrons can hold only 2 atoms, while the second and third energy levels can hold up to 8. Label your atom "Atom B." The style of your model should be similar to the style of Atom A, shown in **Figure 1.**

5. Fill in your data table with the appropriate values for Atom B.

6. On your paper, draw a model of an atom that has an atomic number of 19 and a mass number of 39. Label this atom "Atom C." Fill in the appropriate values for this atom in your data table.

7. On your paper, draw a model of an atom that has an atomic number of 17 and a mass number of 35. Label this atom "Atom D." Fill in the appropriate values for this atom in your data table.

DATA TABLE 1

Atom	# of protons	# of neutrons	Atomic number	Mass number
Atom A				
Atom B				
Atom C				
Atom D				

| Drawing Atomic Models (cont.)

DATA TABLE 2

Atom	Total # of electrons	# of energy levels	# of electrons in highest energy level
Atom A			
Atom B			
Atom C			
Atom D			

PLACING ELEMENTS IN THE PERIODIC TABLE

8. Figure 2 shows part of a simple periodic table that is partially filled in. Copy this table onto a separate sheet of paper. Locate the proper places for Atoms A, B, C, and D on the table. On your copy of the table, write the name of the atom (A, B, C, or D) and the atomic number in the appropriate box for each atom.

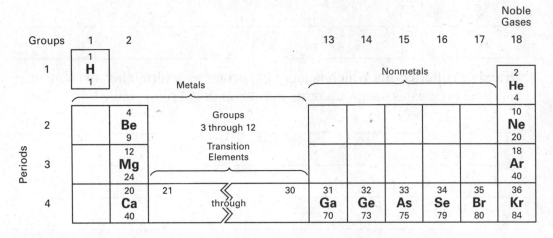

FIGURE 2

ANALYSIS

1. Recognizing patterns What value in the data tables do Atoms A, B, and C have in common? How is this similarity reflected in their positions in the periodic table?

_____

_____

_____

_____

Drawing Atomic Models (cont.)

2. Recognizing patterns How does the period number of each atom compare with the number of energy levels in the atom?

_____

_____

_____

_____

CONCLUSIONS

1. Drawing conclusions Examine a periodic table in your classroom or in your textbook. What are the proper names of Atoms A, B, C, and D?

_____

_____

_____

_____

_____

2. Drawing conclusions Which is more important in determining an element's chemical properties: its group or its period on the periodic table?

_____

_____

_____

_____

3. Applying conclusions Find the element fluorine on a periodic table. How many protons does a fluorine atom have? How many electrons does a neutral fluorine atom have? Which one of the atoms that you studied in this experiment is most chemically similar to fluorine?

_____

_____

_____

_____

_____

Drawing Atomic Models (cont.)

EXTENSIONS

1. **Research and communications** Examine a periodic table. Notice that the second and third periods (the second and third horizontal rows) have a space between the second and third columns. This space allows the transition elements (whose atomic numbers are 21–30) to fit into the table in later periods. Research the number of electrons that can fit into energy levels higher than the third level. Write a short paragraph that describes how electrons fit into these higher levels and that explains why a space must be left in the lower periods on the periodic table.

Name _____ Class _____ Date _____

Extracting Iron from Cereal

Introduction

You have probably walked down the cereal aisle in a supermarket and seen the phrase "Fortified with iron" on many of the boxes. That means iron has been added to the cereal. Why would iron be added to cereal? Iron helps carry oxygen to different parts of your body. Green, leafy vegetables are good sources of iron, but iron-fortified cereals are, too. How could you find out if there is iron in your cereal? One way you can tell is by using a magnet. Iron is found in two forms: as an element and as part of a compound. Elemental iron is attracted to magnets, while iron in a compound form is not. If the cereal contains elemental iron, the iron can be extracted with the magnet. In this experiment, you will attempt to extract iron from cereal and compare the iron extracted from cereal to iron found in iron supplement capsules.

OBJECTIVES

Compare the effects of a magnetic rod on metallic iron and on an iron compound.

Infer whether the iron in iron-fortified cereal is in a compound or in an elemental, metallic state.

MATERIALS

beakers, 500 mL (2)
graduated cylinder, 1 L
iron capsules, dietary (3)
iron-fortified cereal (90 g)
iron filings (small amount)

magnetic stirring rods, each in a small plastic bag (2)
watch or clock
water

SAFETY

- Secure loose clothing and remove dangling jewelry. Don't wear open-toed shoes or sandals in the lab.

- Check the condition of glassware before and after using it. Inform your teacher of any broken, chipped, or cracked glassware because it should not be used.

- Do not pick up broken glass with your bare hands. Place broken glass in a specially designated disposal container.

Extracting Iron from Cereal (cont.)

Procedure

1. Move one of the stirring rods close to the iron filings. Observe what happens to the filings.

2. Remove any filings from the outside of the bag, and discard them.

3. Thoroughly crush the cereal. Place the cereal in the first beaker, and add 100 mL of water. Set the beaker aside. Do not eat the cereal.

4. Empty the contents of the iron capsules into the second beaker, and add 100 mL of water. Set the second beaker aside. Do not eat the iron supplement capsules.

5. Roll up the bags containing the stirring rods so that they are easier to hold. After the cereal has become soggy (10 to 15 minutes), use a stirring rod, magnet side down, to slowly stir the contents of each beaker for 5 minutes.

6. Carefully remove both stirring rods, and examine them closely.

Analysis

1. **Describing events** What happened to the iron filings when the stirring rod approached them?

_____

_____

_____

2. **Describing events** What did you see on the stirring rod from the first beaker?

_____

_____

_____

3. **Describing events** What did you see on the stirring rod from the second beaker?

_____

_____

_____

Extracting Iron from Cereal (cont.)

CONCLUSIONS

1. **Interpreting information** Based on your observations in step 1, are the iron filings made of elemental iron or of an iron compound? Explain your answer.

_____

_____

_____

2. **Drawing conclusions** Based on your observations, do you think elemental iron is present in one or both of the beakers? Explain your answer.

_____

_____

_____

3. **Drawing conclusions** What form of iron was in the iron supplement capsules? Explain your answer.

_____

_____

_____

4. **Applying conclusions** If you dip a magnetic stirring rod into cereal and no iron sticks to the rod, does that mean there is no iron in the cereal? Explain your answer.

_____

_____

_____

EXTENSIONS

1. **Research and communications** Find out what the terms *fortified* and *enriched* mean when they appear on nutritional food labels. Explain.

_____

_____

_____

_____

_____

Combining Elements

Introduction

If you have ever left something metallic out in the rain, you may have returned to find the item spotted with rust. Rust forms when iron combines with oxygen in damp air to make the compound iron(III) oxide. The formation of rust from iron is an example of a synthesis reaction. A synthesis reaction is any reaction in which two or more substances combine to form a single compound. The resulting compound has different chemical and physical properties than the substances that compose the compound do. In this activity, you will synthesize, or create, copper(II) oxide from the elements copper and oxygen.

OBJECTIVES

Describe the change in appearance of copper when copper combines with oxygen to form copper(II) oxide.

Compute the change in mass that occurs in a synthesis reaction.

Explain why mass increases in a synthesis reaction.

MATERIALS

Bunsen burner or portable burner	ring stand and ring
copper powder	spark igniter
evaporating dish	tongs
metric balance	weighing paper
protective gloves	wire gauze

SAFETY

- Never taste, touch, or smell chemicals unless specifically directed to do so.

- Secure loose clothing and remove dangling jewelry. Don't wear open-toed shoes or sandals in the lab.

- Wear an apron or lab coat to protect your clothing when working with chemicals.

- Wear safety goggles when working around chemicals, acids, bases, flames, or heating devices. Contents under pressure may become projectiles and cause serious injury.

- Avoid wearing contact lenses in the lab.

- If any substance gets in your eyes, notify your instructor immediately and flush your eyes with running water for at least 15 minutes.

- Check the condition of glassware before and after using it. Inform your teacher of any broken, chipped, or cracked glassware because it should not be used.

| Combining Elements (cont.)

- In order to avoid burns, wear heat-resistant gloves when handling chemicals.
- If you are unsure of whether an object is hot, do not touch it.
- Avoid wearing hair spray or hair gel on lab days.

Procedure

1. Use the metric balance to measure the mass (to the nearest 0.1 g) of the empty evaporating dish. Record this mass in a table like the one shown below.

2. Place a piece of weighing paper on the metric balance, and measure approximately 10 g of copper powder. Record the mass (to the nearest 0.1 g) in the table. Caution: When working with copper powder, wear protective gloves.

3. Use the weighing paper to place the copper powder in the evaporating dish. Spread the powder over the bottom and up the sides as much as possible. Discard the weighing paper.

4. Set up the ring stand and ring. Place the wire gauze on top of the ring. Carefully place the evaporating dish on the wire gauze.

5. Place the Bunsen burner under the ring and wire gauze. Use the spark igniter to light the Bunsen burner. Caution: When working near an open flame, use extreme care.

6. Heat the evaporating dish for 10 min.

7. Turn off the burner, and allow the evaporating dish to cool for 10 min. Use tongs to remove the evaporating dish, and place it on the balance to determine the mass. Record the mass in your data table.

DATA TABLE

Object	Evaporating dish	Copper powder	Copper + evaporating dish after heating	Copper(II) oxide
Mass (g)				

ANALYSIS

1. Analyzing data Compute the mass of the product of the reaction—copper(II) oxide—by subtracting the mass of the evaporating dish from the mass of the evaporating dish and copper powder after heating. Record this mass in your data table.

Combining Elements (cont.)

2. Examining data Did the mass of the substance in the dish increase or decrease as a result of heating the copper?

_____

_____

_____

3. Describing events What evidence of a chemical reaction did you observe after the copper was heated?

_____

_____

_____

CONCLUSIONS

1. Drawing conclusions Explain why a change in mass occurred as a result of the reaction.

_____

_____

_____

2. Drawing conclusions How does the change in mass support the idea that this reaction is a synthesis reaction?

_____

_____

_____

_____

3. Interpreting information Where did the oxygen in this reaction come from?

_____

_____

_____

Combining Elements *(cont.)*

4. **Evaluating methods** Why was powdered copper rather than a small piece of copper used? (Hint: How does surface area affect the rate of the reaction?)

_____

_____

_____

5. **Evaluating methods** Why was the copper heated?

_____

_____

_____

6. **Applying conclusions** Sometimes, the copper bottoms of cooking pots turn black after the pots are used. How is that similar to the results you obtained in this lab?

_____

_____

_____

EXTENSIONS

1. **Research and communications** Research methods for preventing the formation of rust on automobiles. Explain why it is necessary to paint cars and to patch nicks in the paint. Also, describe at least one additional method for preventing rust on automobiles.

_____

_____

_____

Separating Substances in a Mixture

Introduction

A mixture is a combination of one or more pure substances. A heterogeneous mixture is a mixture in which the substances are not uniformly mixed. Because the substances in a mixture may have different physical properties, these properties can be used to separate the substances from the mixture. In this experiment, you will examine several substances to determine their physical properties. Then you will develop your own methods for extracting these substances from a mixture, and carry out an experiment to test those methods.

OBJECTIVES

Identify key physical properties that could be used to extract a substance from a mixture.

Develop methods for extracting substances from a mixture.

Evaluate the effectiveness of your methods, and recommend improvements.

MATERIALS

balance	nuts, small
beaker, 1 L	oven mitts, (1 pair)
bowl, empty	pepper, cracked
bowl with mixture of substances	plastic-foam cups (6)
craft stick	sand
filter screens (2)	sieve or colander, small
graduated cylinder	spoon, plastic
hot plate	sugar
iron filings	towel, small
magnet, strong	water

SAFETY

- Always use caution when working with chemicals.

- Never taste, touch, or smell chemicals unless specifically directed to do so.

- Secure loose clothing and remove dangling jewelry. Don't wear open-toed shoes or sandals in the lab.

- Wear an apron or lab coat to protect your clothing when working with chemicals.

- If a spill gets on your skin or on your clothing or in your eyes, rinse it immediately, and alert your instructor.

Separating Substances in a Mixture (cont.)

- Wear safety goggles when working around chemicals, acids, bases, flames, or heating devices. Contents under pressure can become projectiles and cause serious injury.

- Avoid wearing contact lenses in the lab.

- If any substance gets in your eyes, notify your instructor immediately and flush your eyes with running water for at least 15 minutes.

- Check the condition of glassware before and after using it. Inform your teacher of any broken, chipped, or cracked glassware because it should not be used.

- Do not pick up broken glass with your bare hands. Place broken glass in a specially designated disposal container.

- In order to avoid burns, wear heat-resistant gloves when handling chemicals.

- If you are unsure whether an object is hot, do not touch it.

- Avoid wearing hair spray or hair gel on lab days.

- Dispose of all sharps (broken glass and other contaminated sharp objects) and other contaminated materials (biological and chemical) in special containers as directed by your instructor.

Procedure

1. Learning the physical properties of a substance can help you decide how to separate it from other substances. Observe the physical properties of each of the following substances: cracked pepper, iron filings, nuts, sand, sugar, and water. Caution: Do not taste or eat any of these substances.

2. For each substance, answer the following questions in the Physical properties column of the chart on the next page: Does this substance dissolve in water? Does it float? Is it magnetic? Are its pieces large or small relative to the pieces of other substances? Is it a solid, a liquid, or a gas?

3. Look over the properties you recorded in step 2. For each substance, determine which characteristic would help you best distinguish it from the other substances. Choose from among the following properties: size, shape, density, state of matter, solubility, and magnetic attraction. Record a distinguishing characteristic on the chart for each substance.

Separating Substances in a Mixture (cont.)

DATA TABLE Separation of Substances

Substance	Make Observations		Form Hypotheses	Conduct an Experiment	
	Physical properties	Distinguishing characteristic	Method of separation	Check when done	Mass extracted (g)
Pepper	small, but larger than sand; does not dissolve in water; floats; not magnetic	density	Allow the mixture to settle. Scrape the layer of floating pepper from the water with a craft stick.		
Nuts					
Sand					
Iron filings					
Water					
Sugar					

Separating Substances in a Mixture (cont.)

4. You have been given a bowl containing a mixture of all the substances you have observed: pepper, iron filings, nuts, sand, sugar, and water. Formulate hypotheses about how you could separate each substance from the rest of the mixture using the materials provided for you in the lab. It may help to look over the materials list. If you are working in a group, discuss your proposed methods with the others in the group and decide together on a method for each substance. Describe your proposed methods in the "Method of separation" column in the table. The first row in the chart is filled in to help you get started.

5. Have your teacher approve the methods you have described, then conduct an experiment to test your methods.

6. Measure the mass of an empty cup, and record the mass.

7. Follow your plans to separate each substance from the mixture. If your plans change, be sure to note the changes in the method column on the chart. Be careful not to spill any of the substances on the floor as you work. Protect your hands with oven mitts when working with the hot plate, and be careful not to burn any of the substances. Never place any of the substances directly on the hot plate. As you extract the substances from the mixture, store each substance in a different cup, and label each cup.

8. When you have finished extracting all the substances from the mixture, measure the mass of each cup and its contents. For each measured mass, subtract the mass of the empty cup that you measured in step 6, and record your results in the last column on the chart.

ANALYSIS

1. **Organizing data** List the substances in the mixture in order from the least mass extracted to the greatest mass extracted.

   _____

   _____

   _____

2. **Classifying** Was this mixture a homogeneous mixture or a heterogeneous mixture? Was it a suspension, a colloid, an emulsion, or a solution?

   _____

   _____

   _____

Separating Substances in a Mixture (cont.)

CONCLUSIONS

1. Evaluating results Were your measurements of the masses of the substances in the mixture accurate? Why or why not?

_____

_____

_____

_____

_____

_____

_____

2. Evaluating methods How could you improve your methods to better separate each substance?

_____

_____

_____

_____

_____

_____

_____

EXTENSIONS

1. Research and communications A centrifuge is a device that uses centrifugal force to separate substances of different masses. Centrifuges are often used by biologists and biochemists to separate parts of cells or biological molecules of different masses. Research how centrifuges work, and write a short paragraph explaining what you learn.

Skills Practice Lab

Modeling Radioactive Decay with Pennies

Introduction

Imagine existing more than 5000 years and still having more than 5000 to go! That is exactly what the unstable element carbon-14 does. Carbon-14 is an unstable isotope of carbon. Carbon-14 is used in the radioactive dating of material that was once alive, such as fossil bones. Every 5730 years, half of the carbon-14 in a fossil specimen decays or breaks down into the stable element nitrogen-14. In the following experiment you will see how pennies can be a model for the same kind of decay.

OBJECTIVES

Discover how the number of coins remaining after shaking, pouring, and selecting for tails changes with each shake.

Graph the number of coins remaining as a function of the number of shakes.

Compare the graph of the number of coins remaining to a graph of radioactive decay.

MATERIALS

containers with covers, large (2)   pennies (100)

SAFETY

• Secure loose clothing and remove dangling jewelry. Don't wear open-toed shoes or sandals in the lab.

Procedure

1. Place 100 pennies in a large, covered container. Shake the container several times and remove the cover. Carefully empty the container on a flat surface, and make sure the pennies don't roll away.

2. Remove all the coins that have the head side of the coin turned upward, and place them in a separate container. In a data table like the one on the next page, record the number of pennies removed and the number of pennies remaining.

Modeling Radioactive Decay with Pennies (cont.)

3. Place the remaining pennies (with the tail side showing) back into the original container. Shake the container, and empty it onto the flat surface again. Sort out the pennies, and record data as in step 2. Remember to remove only the coins showing heads. Repeat this process until no pennies are left in the container. If the process requires more than nine steps, extend your data table as needed.

4. When you are finished, return all pennies to the original container, and clean up your work area.

DATA TABLE

Shake number	Number of coins remaining	Number of coins removed
1		
2		
3		
4		
5		
6		
7		
8		
9		

ANALYSIS

1. **Constructing graphs** Draw a graph to plot your data. Label the x-axis "shake number," and label the y-axis "Pennies remaining." Using data from your data table, plot the number of coins remaining after each shake.

_____

_____

_____

2. **Examining data** How many shakes did it take before the number of pennies remaining was about one half the original number of pennies (about 50)? How many shakes did it take before the number of pennies remaining was about one-fourth the original number of pennies (about 25)?

_____

_____

_____

Modeling Radioactive Decay with Pennies (cont.)

CONCLUSIONS

1. **Drawing conclusions** What is the half-life of the pennies in this experiment, in numbers of shakes?

   _____

   _____

   _____

2. **Analyzing graphs** Examine the graph below. Compare the graph you have made for pennies with the graph for carbon-14. Explain any similarities that you see between the graphs.

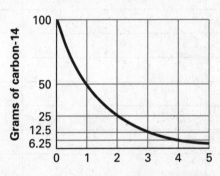

Half-Life of Carbon-14

Number of half-lives (5730 y)

   _____

   _____

   _____

3. **Analyzing conclusions** Imagine that you have found a fossilized leg bone of some unknown mammal. Based on the size of the bone, you determine that it should have contained about 100 g of carbon-14 when the animal was alive. The bone now contains about 12.5 g of carbon-14. How old is the bone?

   _____

   _____

   _____

Modeling Radioactive Decay with Pennies (cont.)

EXTENSIONS

1. Research and communications Carbon-14 is used to date materials as old as about 60000 years. However, the age of Earth is thought to be 4.5 billion years, and life is thought to have existed on Earth for close to 4 billion years. These dates have been determined in part using radioactive dating methods. Research radioactive dating methods that can measure ages much older than 60000 years. Write a paragraph explaining at least two such methods. Include the names of the radioactive elements, the names of the elements they decay into, and the half-life of the reactions.

Name _____ Class _____ Date _____

Skills Practice Lab

Testing Reaction Time

Introduction

Objects in free fall near Earth's surface accelerate under the influence of gravity. This acceleration is constant, but it may be reduced somewhat by air resistance. Near Earth's surface, the acceleration due to gravity is 9.8 m/s^2. You can use this acceleration and a measurement of falling distance to calculate the time that an object is in free fall.

Predicting human reaction time is more complicated than predicting a simple physical event such as the time an object takes to fall. Human reaction time depends on many factors, such as the time a signal takes to go from your eyes to your brain, the time your brain takes to process the signal, and the time your brain takes to signal a reaction. Although the steps in reaction time are complicated, you can determine overall reaction time by using simple measurements.

In this lab, you will test your reaction time by measuring how far a meterstick falls before you can catch it. You will measure the distance that the meterstick falls and will then use an equation based on free-fall acceleration to determine the amount of time the meterstick took to fall. You will also measure the amount of time the meterstick falls by using a stopwatch and will compare the measured reaction time with the calculated reaction time.

OBJECTIVES

Calculate reaction time of a falling object by using falling-distance data.

Compare reaction times measured by using a stopwatch with reaction times calculated from falling-distance data.

Evaluate the methods used to determine reaction time in the experiment.

MATERIALS

calculator   paper
meterstick   stopwatch

SAFETY

- Secure loose clothing and remove dangling jewelry. Don't wear open-toed shoes or sandals in the lab.

Procedure

1. Have each person in the group prepare a data table like the one on the next page. Each person should write his or her name at the top of the table.

2. Work in pairs. Hold the meterstick vertically with the zero end down. Have your partner stand in front of you with his or her thumb and forefinger open about an inch and a half apart and lined up with the bottom (zero end) of the meterstick.

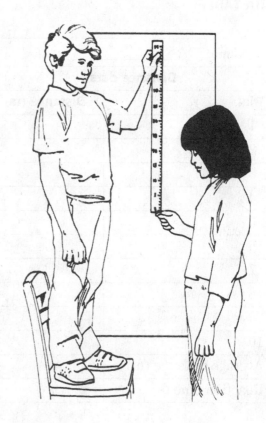

3. Drop the meterstick. Your partner should try to catch the meterstick between his or her thumb and forefinger as quickly as possible. Once your partner catches the meterstick, you can measure how far the meterstick fell by reading the point on the meterstick that your partner grabbed. Convert this distance to meters, and record it in your data table.

4. Repeat steps 2–3 nine more times, for a total of 10 trials. Record all falling distances in your data table.

5. Repeat steps 2–3 again, but this time use a stopwatch. Try to start the stopwatch at the same time that you release the meterstick, and stop the stopwatch as soon as your partner catches the meterstick. Repeat this process until you have completed 5 trials. Record all times in the lower part of your data table. You do not need to measure or record distance for these time trials.

6. Change places with your partner, and repeat steps 2–5. Record all falling distances in another data table. If you are working in a group of more than two people, continue repeating steps 2–5 until everyone has had a chance to test reaction time.

| Testing Reaction Time (cont.)

DATA TABLE

Name: _____			
Distance data		**Time data**	
Trial	Distance (m)	Trial	Time (s)
1		11	
2		12	
3		13	
4		14	
5		15	
6		Average time (s)	
7			
8			
9			
10			
Average distance (m)			
Reaction time (s)			

ANALYSIS

1. **Organizing data** For each person in the group, calculate the average falling distance by adding the distance for the first 10 trials and then dividing by 10. Record the average in your data table.

_____

_____

_____

2. **Organizing data** For each person in the group, calculate reaction time from the average distance by using the following formula:

$$reaction\ time = \sqrt{\frac{2 \times average\ distance}{9.8\ m/s^2}}$$

_____

_____

_____

Testing Reaction Time (cont.)

3. Organizing data For each person in the group, calculate the average falling time by adding the time for the last 5 trials and then dividing by 5. Record the average in your data table.

_____

_____

_____

CONCLUSIONS

1. Evaluating methods How does the average falling time you calculated by using data from the stopwatch compare with the reaction time you calculated by using the average falling distance? Should these values be the same? Why might these values be different?

_____

_____

_____

_____

_____

2. Evaluating results How does your reaction time compare with those of other members of your group? Who has the fastest reaction time?

_____

_____

_____

_____

3. Evaluating data How did your reaction time change over the first 10 trials? Do you see any evidence of improvement with practice?

_____

_____

_____

_____

_____

4. Evaluating methods Why did each of you do 10 trials to find an average distance instead of doing just 1 trial? Do you think your results would be more accurate if you did even more trials? Explain your answer.

_____

_____

_____

_____

_____

EXTENSIONS

1. Designing experiments Repeat this experiment, but blindfold the person who is catching the meterstick. As soon as you drop the meterstick, signal your blindfolded partner by using your voice. Repeat this process for several trials. Then, repeat the process for several more trials, but signal your blindfolded partner by tapping your partner on the arm or shoulder. You do not need to use the stopwatch; use only falling-distance data. Which of these signaling methods produces the fastest reaction time? How do these reaction times compare with the reaction time when the catcher is not blindfolded?

_____

_____

_____

_____

_____

Skills Practice Lab

Exploring Work and Energy

Introduction

Common sense may tell you that pushing or pulling a heavy object up a ramp is easier than lifting the object straight up. Pushing or pulling an object up a ramp usually does require less force than lifting it does, but these actions do not always require less work. Work in the scientific sense is calculated as the force applied to an object times the distance the object moves. When you push or pull an object up a ramp, you increase the total distance that you have to move the object in order to raise it to a given height. Furthermore, you have to overcome the force of friction between the object and the ramp.

In this experiment, you will measure the forces required to move objects across a level surface, to lift the objects straight up, and to pull the objects up an inclined plane. You will also calculate the work done on the objects in these three cases. Then, you will compare the amount of force and the amount of work required in each case.

OBJECTIVES

Measure the force required to move a mass over a certain distance.

Compute the work done on a mass.

Compare the work done on a mass and the force required to move the mass by using different methods.

MATERIALS

clamps   inclined plane
cord, 1.00 m   masking tape
force meters, spring scales (2)   meterstick
hooked masses, one set   stopwatch

Safety

• Secure loose clothing and remove dangling jewelry. Don't wear open-toed shoes or sandals in the lab.

Exploring Work and Energy (cont.)

Procedure
PULLING MASSES

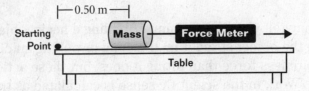

1. At one edge of the tabletop, place a tape mark to represent a starting point. From this mark, measure exactly 0.25 m and 0.50 m. Place a tape mark at each measured distance.

2. Securely attach the 1 kg mass to one end of the cord and the force meter to the other end. The force meter will measure the force required to move the mass through different displacements.

   Set up the apparatus, and attach all masses securely. Perform this experiment in a clear area. Swinging or dropped masses can cause serious injury.

3. Place the mass on the table at the starting point. Hold the force meter parallel to the tabletop so that the cord is taut between the force meter and the mass. Carefully pull the mass at a slow, constant speed along the surface of the table to the 0.25 m mark (this may require some practice). As you pull, observe the force measured on the force meter.

4. Record the force and distance in a table using the appropriate SI units (newtons and meters).

5. Repeat steps 3 and 4 for a distance of 0.50 m.

6. Repeat steps 3–5 with a 0.2 kg mass.

LIFTING MASSES

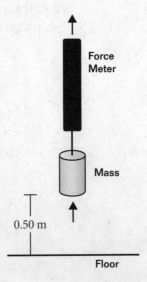

7. Using masking tape, secure a meterstick vertically against the wall so that the 0.00 m end touches the floor.

Exploring Work and Energy (cont.)

8. Securely attach the 1 kg mass to one end of the cord and the force meter to the other end.

9. Place the mass on the floor beside the meterstick. Hold the force meter parallel to the wall so that the cord is taut between the force meter and the mass. Carefully lift the mass vertically at a slow, constant speed to the 0.25 m mark on the meterstick. Be sure that the mass does not touch the wall during any part of the process. As you lift, observe the force measured on the force meter. Be careful not to drop the mass.

10. Record the force and distance in your notebook using the appropriate SI units.

11. Repeat steps 9 and 10 for a vertical distance of 0.50 m.

12. Replace the 1 kg mass with the 0.2 kg mass, and repeat steps 9–11.

DISPLACING MASSES ON AN INCLINED PLANE

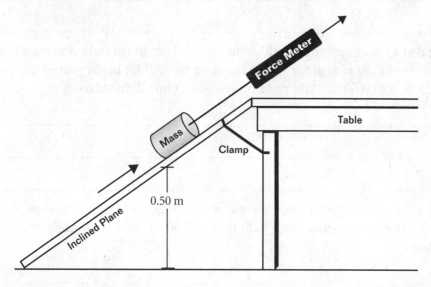

13. Carefully clamp an inclined plane to the tabletop so that the base of the inclined plane rests on the floor. Make sure the inclined plane is in a location where it will not obstruct traffic or block aisles or exits.

14. Measure vertical distances of 0.25 m and 0.50 m above the level of the floor. Use masking tape to mark each level on the inclined plane. Also measure the distance along the inclined plane to each mark. Record all distances in your notebook using the appropriate SI units. Be sure to label the vertical distance and the distance along the inclined plane.

15. Attach the 1 kg mass to the lower end of the cord and the force meter to the other end.

16. Place the mass at the base of the inclined plane. Hold the force meter parallel to the inclined plane so that the cord is taut between the force meter and the mass. Carefully pull the force meter at a slow, constant speed parallel to the surface of the inclined plane until the mass has reached the vertical 0.25 m

Exploring Work and Energy (cont.)

mark on the inclined plane. As you pull, observe the force measured on the force meter.

17. Using the appropriate SI units, record the force and distance in your notebook.

18. Repeat steps 16 and 17 for a vertical distance of 0.50 m.

19. Repeat steps 16–18 for the 0.2 kg mass.

ANALYSIS

1. Organizing data Calculate the work done on the 1 kg mass pulled across the table a distance of 0.25 m. Use the work equation, $W = F \times d$. Add your answer to your data table.

_____

_____

_____

2. Organizing data Calculate the work done on the 1 kg mass pulled across the table a distance of 0.50 m and the work done on the 0.2 kg mass pulled distances of 0.25 m and 0.50 m. Add your answers to your data table.

_____

_____

_____

3. Organizing data Calculate the work done on the 1 kg and 0.2 kg masses lifted vertically through distances of 0.25 m and 0.50 m. Add your answers to your data table.

_____

_____

_____

4. Organizing data Calculate the work done on the 1 kg and 0.2 kg masses pulled up the inclined plane through vertical distances of 0.25 m and 0.50 m. In each calculation, remember to use the distance that the mass was pulled along the inclined plane, not the vertical distance. Add your answers to your data table.

_____

_____

_____

Exploring Work and Energy (cont.)

5. Analyzing data In each of the three setups, did you exert the same force on the 1 kg mass as you did on the 0.2 kg mass to move them an equal distance?

_____

_____

_____

6. Analyzing data In each of the three setups, did moving the mass 0.50 m require more force than moving the same mass 0.25 m did?

_____

_____

_____

7. Analyzing results In each of the three setups, did you do more work on the 1 kg mass or on the 0.2 kg mass to move the masses a distance of 0.50 m?

_____

_____

_____

8. Analyzing results In each of the three setups, did you do more work on the 1 kg mass when you moved it a distance of 0.25 m or when you moved it a distance of 0.5 m?

_____

_____

_____

CONCLUSIONS

1. Interpreting information What force did you pull against when you pulled the masses horizontally across the table?

_____

_____

_____

2. **Interpreting information** What force did you pull against when you lifted the masses vertically?

_____

_____

_____

3. **Interpreting information** What forces did you pull against when you pulled the masses up the inclined plane?

_____

_____

4. **Drawing conclusions** Compare your data on pulling the 1 kg mass across the table to your data on lifting the 1 kg mass vertically. Which force was greater, the force of friction between the mass and the table or the force of gravity acting on the mass?

_____

_____

_____

5. **Drawing conclusions** Compare your data on lifting the 1 kg mass vertically through 0.50 m to your data on pulling the 1 kg mass up the inclined plane to a height of 0.50 m. In which case did you exert a greater force, and in which case did you do more work on the mass?

_____

_____

_____

6. **Defending conclusions** You should have found that you did more work on the mass to pull it up the inclined plane to a vertical height of 0.50 m than you did to lift it vertically to the same height (if you did not get that result, check your calculations or ask your teacher to look over your data). Explain why this is the case.

_____

_____

_____

_____

Exploring Work and Energy (cont.)

7. Making predictions How could you adjust the inclined plane so that moving the mass through the same vertical displacement would require less force?

_____

_____

_____

8. Applying conclusions Would decreasing the angle of the ramp to the floor increase or decrease the amount of work done to move the mass to a vertical height of 0.50 m? Explain your answer.

_____

_____

_____

EXTENSIONS

1. Designing experiments Design an experiment to test the amount of work done and the amount of force required to pull a mass up an inclined plane at different angles to the ground. Make a hypothesis about how changing the angle will affect the work done and the force required to lift the mass to a given vertical height.

Skills Practice Lab
Energy Transfer and Specific Heat

Introduction

Heat is the energy transfer between objects that have different temperatures. Energy moves from objects that have higher temperatures to objects that have lower temperatures. If two objects are left in contact for a while, the warmer object will cool down and the cooler object will warm up until they eventually reach the same temperature. In this activity, you will combine equal masses of iron nails and water that have different temperatures to determine whether the iron nails or the water has a greater effect on the final temperature.

OBJECTIVES

Predict the final temperature when equal amounts of iron and water that have different temperatures are combined.

Measure the initial temperature of the iron, the initial temperature of the water, and the final temperature of the combined iron and water.

Compare the results of the experiment to the prediction.

MATERIALS

graduated cylinder, 100 mL

marker

metric balance

nails (10–12)

paper towels

plastic-foam cups, 9 oz (2)

rubber band

string, 30 cm long

thermometer

water, cold

water, hot

SAFETY

- Secure loose clothing and remove dangling jewelry. Don't wear open-toed shoes or sandals in the lab.

- In order to avoid burns, wear heat-resistant gloves when instructed to do so.

- If you are unsure of whether an object is hot, do not touch it.

- Check the condition of glassware before and after using it. Inform your teacher of any broken, chipped, or cracked glassware because it should not be used.

- Do not pick up broken glass with your bare hands. Place broken glass in a specially designated disposal container.

Energy Transfer and Specific Heat (cont.)

Procedure
MAKE A PREDICTION

1. When you combine equal amounts of iron and water, each of which has a different temperature, will the final temperature be closer to the initial temperature of the iron, closer to the initial temperature of the water, or halfway in between?

CONDUCT AN EXPERIMENT

2. Use a rubber band to bundle the nails together. Measure and record the mass of the bundle in a data table like the one below. Tie a length of string around the bundle, leaving one end of the string 15 cm long. Use the marker to label the cups "A" and "B".

3. Put the bundle of nails into cup A. Let the end of the string hang over the side of the cup. Fill the cup with enough hot water to cover the nails, and set the cup aside for at least 5 min.

4. Use the graduated cylinder to measure the amount of cold water that has a mass equal to the mass of the nails (1 mL of water has a mass of 1 g). Record this volume in your data table. Pour the water from the graduated cylinder into cup B.

5. Measure and record the temperature of hot water that covers the nails in cup A and the temperature of the water in cup B. The temperature of the water in cup A represents the temperature of the nails.

6. Use the string to transfer the bundle of nails to the water in cup B. Use the thermometer to monitor the temperature of the water in cup B. When the temperature in cup B stops changing, record this temperature in the "Final temperature" column of your data table.

7. For Trial 2, repeat steps 3–6, but this time use cold water in step 3 and hot water in step 4. Start step 3 by filling cup A with cold water. In step 4, measure hot water to pour into cup B. Record all your measurements.

8. Empty the cups and dry the nails.

DATA TABLE

Trial	Mass of nails (g)	Volume of water (mL)	Initial temp of water and nails (°C)	Initial temp of water without nails (°C)	Final temp of water and nails combined (°C)
1					
2					

Energy Transfer and Specific Heat (cont.)

1. **Examining data** In Trial 1, you used equal masses of cold water and nails. Did the final temperature support your prediction? Explain.

_____

_____

_____

2. **Examining data** In Trial 2, you used equal masses of hot water and nails. Did the final temperature support your prediction? Explain.

_____

_____

_____

3. **Analyzing data** In Trial 1 and Trial 2, which material—the water or the nails—changed temperature the most after you transferred the nails? Explain your answers.

_____

_____

_____

CONCLUSIONS

1. **Drawing conclusions** In Trial 1, the cold water gained energy. Where did the energy come from?

_____

_____

_____

2. **Drawing conclusions** How does the energy gained by the nails in Trial 2 compare with the energy lost by the hot water in Trial 2? Explain.

_____

_____

_____

_____

_____

3. Drawing conclusions Which material seems to be able to hold energy better? Explain your answer.

_____

_____

_____

4. Drawing conclusions Specific heat is a property of matter that tells how much energy is required to change the temperature of 1 kg of a material by 1 °C. Which material, iron or water, used in this activity has a higher specific heat (changes temperature less for a given amount of energy)?

_____

_____

_____

5. Applying conclusions Would it be better to have pots and pans made from a material with a high specific heat or a low specific heat? Explain your answer. (Hint: Do you want the pan or the food in the pan to absorb all the energy from the stove?)

_____

_____

_____

6. Evaluating results Share your results with your classmates. Discuss how you would change your prediction to include your knowledge of specific heat.

_____

_____

_____

EXTENSIONS

1. Designing experiments Design an experiment to compare the specific heat of nails made of iron with the specific heat of nails made of a metallic alloy.

Skills Practice Lab

Boyle's Law

Introduction

According to Boyle's law, all gases behave the same when compressed; that is, as increasing pressure is applied to a gas in a closed container, the volume of the gas decreases. Boyle's law may be stated mathematically as $P \sim \dfrac{1}{V}$ or $PV = k$ (where k is a constant). Notice that for Boyle's law to apply, two variables that affect gas behavior must be held constant: the amount of gas and the temperature.

   In this experiment, you will vary the pressure of air contained in a syringe and measure the corresponding change in volume. Because it is often impossible to determine relationships by just looking at data in a table, you will plot graphs of your results to see how the variables are related. You will make two graphs, one of volume versus pressure and another of the inverse of volume versus pressure. From your graphs you can infer the mathematical relationship between pressure and volume in order to verify Boyle's law.

OBJECTIVES

Determine the volume of a gas in a container under various pressures.

Graph pressure-volume data to discover how the variables are related.

Interpret graphs, and verify Boyle's law.

MATERIALS

Boyle's law apparatus
carpet thread

objects of equal mass, approximately
500 g each (4)

SAFETY

- Secure loose clothing and remove dangling jewelry. Don't wear open-toed shoes or sandals in the lab.

- Wear safety goggles when working around chemicals, acids, bases, flames, or heating devices. Contents under pressure may become projectiles and cause serious injury.

- If any substance gets in your eyes, notify your instructor immediately and flush your eyes with running water for at least 15 minutes.

Boyle's Law (cont.)

Procedure

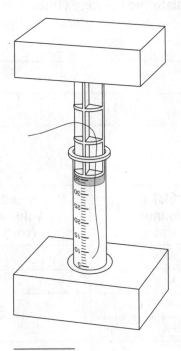

FIGURE A

1. Adjust the piston head of the Boyle's law apparatus so that it reads between 30 and 35 cm³. To adjust, pull the piston head all the way out of the syringe, insert a piece of carpet thread into the barrel, and position the piston head at the desired location, as shown in the figure above.

   Note: Depending upon the Boyle's law apparatus that you use, you may find the volume scale on the syringe abbreviated in cc or cm³. Both abbreviations stand for cubic centimeters (1 cubic centimeter is equal to one milliliter). The apparatus shown in the figure is marked in cm³.

2. While holding the piston in place, carefully remove the thread. Twist the piston several times to allow the head to overcome any frictional forces. Read the volume to the nearest 0.1 cm³. Record this value in your data table as the initial volume for zero masses.

3. Place one of the masses on the piston. Give the piston several twists to overcome any frictional forces. When the piston comes to rest, read and record the volume to the nearest 0.1 cm³.

4. Repeat step 3 for two, three, and four masses, and record your results.

5. Repeat steps 3 and 4 for at least two more trials. Record your results.

6. Clean all apparatus and your lab station at the end of this experiment. Return equipment to its proper place. Ask your teacher how to dispose of any waste materials.

Boyle's Law (cont.)

ANALYSIS

1. **Organizing data** Calculate the average volume of the three trials for masses 0–4. Record your results in your calculations table.

_____

_____

_____

DATA TABLE

Pressure (number of weights)	Trial 1 Volume (cm³)	Trial 2 Volume (cm³)	Trial 3 Volume (cm³)
0			
1			
2			
3			
4			

2. **Organizing data** Calculate the inverse for each of the average volumes. For example, if the average volume for three masses is 26.5 cm³, then 1/volume = 1/26.5/cm³ = 0.0377/cm³.

_____

_____

_____

CALCULATIONS TABLE

Pressure (number of weights)	Average Volume (cm³)	1/Volume (×10²/cm³)
0		
1		
2		
3		
4		

3. Constructing graphs Plot a graph of volume versus pressure. Because the number of masses added to the piston is directly proportional to the pressure applied to the gas, you can use the number of masses to represent the pressure. Notice that pressure is plotted on the horizontal axis and volume is plotted on the vertical axis. Draw the smoothest curve that goes through most of the points.

_____

_____

_____

Volume Versus Pressure

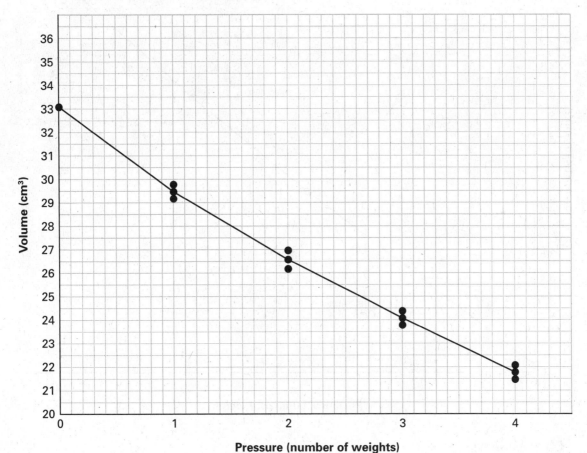

4. **Constructing graphs** Plot a graph of 1/volume versus pressure. Notice that pressure is on the horizontal axis and 1/volume is on the vertical axis. Draw the best line that goes through the majority of the points.

_____

_____

_____

1/V Versus Pressure

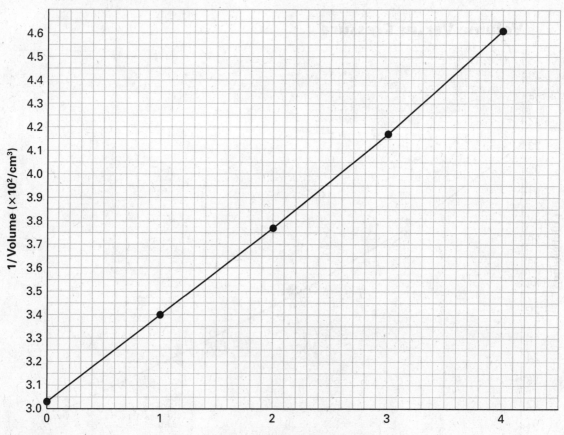

CONCLUSIONS

1. **Analyzing graphs** Do your graphs indicate that a change in volume is directly proportional to a change in pressure? Explain your answer.

_____

_____

_____

Boyle's Law (cont.)

2. Analyzing graphs Based on your graphs, what do you conclude about the mathematical relationship between the pressure applied to a gas and the corresponding volume of the gas?

_____

_____

_____

_____

3. Making predictions Use your graph to predict the volume of the gas if 2.5 masses were used.

_____

_____

_____

_____

4. Defending conclusions Use the kinetic theory description of a gas to explain the observed relationship between the pressure and volume of a gas.

_____

_____

_____

_____

5. Applying conclusions If a normal sea-level recipe is used to prepare a cake at a location 1000 meters below sea level, the cake will be much flatter than expected. Explain why, and offer a solution. (Hint: Consider how barometric pressure differs at this altitude and what effect that may have on the ability of the cake to rise.)

_____

_____

_____

_____

Boyle's Law (cont.)

Extensions

1. **Research and communications** Use the Internet or a library to look up another fundamental gas law, such as Charles's law or Avogadro's law. Write down the mathematical expression of the law. State which quantities the law assumes to be held constant and which quantities may change.

Name _____ Class _____ Date _____

Creating and Measuring Standing Waves

Introduction

When you fix a rope at one end and then shake the other end up and down, waves travel down the rope. You can vary the frequency and wavelength of the waves by changing the rate at which you shake the rope. If the waves are of certain wavelengths, the waves will appear to stop traveling along the rope, and certain points on the rope, called *nodes*, will be completely motionless. Waves that behave this way are called *standing waves*. In this experiment, you will create several different standing waves on a rope. You will calculate the frequency, wavelength, and wave speed for each standing wave and then will look for patterns in these values.

OBJECTIVES

Create standing waves on a rope fixed at one end.

Calculate frequency, wavelength, and wave speed of standing waves.

Compare frequency, wavelength, and wave speed of different standing waves.

MATERIALS

clothesline rope, 3 m long
meterstick
stopwatch

Safety

• Secure loose clothing and remove dangling jewelry. Don't wear open-toed shoes or sandals in the lab.

Creating and Measuring Standing Waves (cont.)

Procedure
PULLING MASSES

1. Tie one end of the rope to a fixed support, such as a doorknob. Hold the free end of the rope, and measure the length of the rope from your hand to the knot at the fixed end. Record the length in a data table like the one shown below.

2. Shake the free end of the rope up and down, starting slowly. Be sure not to move the rope side to side or in a circle. Adjust the frequency until the entire rope is vibrating up and down as a whole (making one "loop," even though the loop is confined to a plane). Try to keep the rope moving like this while minimizing the motion of your hand. There should now be a node at each end of the rope and an antinode in the middle.

3. Using a stopwatch, time how long it takes to shake the rope up and down through 10 cycles (one up-and-down motion is one cycle). Divide this total time by 10 to calculate the frequency of the standing wave. Record the frequency in your data table.

4. Increase the frequency by shaking the free end of the rope up and down faster. Continue to increase the frequency until there is a node in the middle of the rope. Now the rope should have two "loops," with two antinodes.

5. Calculate the frequency of this standing wave with two loops by using the same method as in step 3. Record the frequency in your data table.

6. Increase the frequency again until you have a standing wave with three antinodes (three loops). Determine the frequency as in step 3, and record the frequency in your data table.

DATA TABLE

Total length of rope: _____ m			
Number of antinodes	Frequency (Hz)	Wavelength (m)	Wave speed (m/s)
1			
2			
3			

Creating and Measuring Standing Waves (cont.)

ANALYSIS

1. **Organizing data** Nodes in a standing wave are always one-half of a wave-length apart. Therefore, with one loop in the rope, the rope is one-half of a wavelength long. Calculate the total wavelength by multiplying the length of the rope by 2. Record this wavelength in your data table.

_____

_____

_____

2. **Organizing data** With two loops in the rope, the rope is one wavelength long. With three loops, the rope is one-and-a-half wavelengths long. Calculate the wavelength for these two cases, and record the wavelengths in your data table.

_____

_____

_____

_____

3. **Recognizing patterns** As the frequency increased, did the wavelength increase or decrease?

_____

_____

_____

_____

4. **Organizing data** Once you know the frequency and wavelength of a wave, you can calculate the speed of the wave by using the wave-speed equation:

$$wave\ speed = frequency \times wavelength.$$

Calculate the wave speed for each of the three standing waves that you produced.

_____

_____

_____

_____

Creating and Measuring Standing Waves (cont.)

CONCLUSIONS

1. Drawing conclusions How are the wave speeds of the three different standing waves related?

_____

_____

_____

_____

2. Drawing conclusions How are the frequencies of the three different standing waves related?

_____

_____

_____

_____

3. Making predictions What would the frequency, wavelength, and wave speed of a standing wave with four antinodes on your rope be?

_____

_____

_____

_____

4. Making predictions How could you change the experimental setup to produce standing waves that have a different wave speed?

_____

_____

_____

_____

EXTENSIONS

1. Designing experiments Design an experiment to determine how the speed of standing waves is related to the density (mass per unit length) of the rope used. The results of the experiment should include a graph. If you have time and your teacher approves, carry out your experiment by using at least three kinds of rope.

Skills Practice Lab

Mirror Images

Introduction

Mirrors form images by reflecting light coming from objects. When you view an image in a mirror with your eyes, you are seeing a virtual image, which usually appears to exist somewhere behind the mirror. Depending on the shape of the surface of the mirror, this mirror image may be reversed left to right, turned upside down, or magnified to look larger or smaller. In this lab, you will observe images in a flat mirror, a convex mirror (curved outward), and a concave mirror (curved inward). You will also measure various distances and angles to learn more about the properties of the images formed by these different types of mirrors.

OBJECTIVES

Describe the images of objects in various kinds of mirrors.

Draw diagrams showing paths of light rays reflected from mirrors.

Compare tlight path lengths and angles of incidence and reflection.

MATERIALS

eye charts, both normal and reverse   pencil
meterstick   protractor
mirror, curved   ruler or straightedge
mirror, small and flat   tape
mirror supports   T-pin with pencil eraser
paper   white paper

SAFETY

- Secure loose clothing and remove dangling jewelry. Don't wear open-toed shoes or sandals in the lab.

- Never look directly at the sun through any optical device or use direct sunlight to illuminate a microscope.

- If any substance gets in your eyes, notify your instructor immediately and flush your eyes with running water for at least 15 minutes.

- Check the condition of glassware before and after using it. Inform your teacher of any broken, chipped, or cracked glassware because it should not be used.

- Do not pick up broken glass with your bare hands. Place broken glass in a specially designated disposal container.

| *Mirror Images (cont.)*

Procedure
VIRTUAL IMAGES

1. Secure the normal eye chart to the wall by using strong tape.

2. Choose any line on the chart, and step back until the line can no longer be read clearly. Use masking tape to mark the position on the floor where you are standing. Label this position "reading point."

3. Measure the distance from the eye chart to the reading point with a meter-stick. Record this distance in your notebook, using the appropriate SI units (meters). Also, record the number of the line that you were trying to read.

4. Secure a small, flat mirror against the wall at chest level by using strong tape.

5. Place the back of the reverse eye chart against your chest. Position the chart so that the line that you read appears in the mirror. Step back from the mirror, and hold the eye chart against your chest until the image of this line is barely readable.

6. Use masking tape to mark the position on the floor where you are standing. Label this position "new point."

7. Measure the distance from the eye chart to the new point. Record this distance in your notebook, using the appropriate SI units.

FLAT MIRRORS

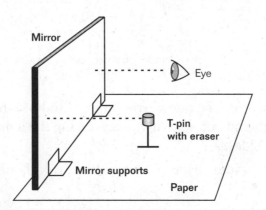

8. Using two mirror supports, vertically stand one flat mirror on a table, away from the edge, as shown in the figure above. Place a sheet of white paper on the tabletop so that the front of the mirror faces the paper. Tape the paper and mirror supports to the table so that they do not slide.

9. Remove the eraser from a pencil. Secure the eraser on the tip of the T-pin to cover the point. Using tape, carefully secure the T-pin on the tabletop, with the T side down, in front of the mirror. The point with the eraser should be pointing up.

10. Wearing a pair of safety goggles, move your head to one side of the pin. Close one eye, and place your open eye at the level of the tabletop. Observe the image of the pin in the mirror.

Mirror Images (cont.)

11. Use a ruler to draw a straight line on the paper from the image of the pin in the mirror to the position of your eye. Label this line "outgoing beam." Use a ruler to draw a straight line from the object to the mirror's surface. Label this line "incoming beam." Both lines should meet at the same point on the mirror's surface.

12. Draw a line on the paper from the position of your eye perpendicular to the mirror's surface. Draw a line from the object perpendicular to the mirror's surface. Both lines should be parallel to each other. These lines will form angles with the lines you drew in step 11.

13. Measure the angle between the line labeled "outgoing beam" and the nearest perpendicular line. Measure the angle between the line labeled "incoming beam" and the nearest perpendicular line. Record these angles in your notebook, using units of degrees.

14. Move your eye to a new position. Repeat steps 10–13.

15. Move your eye to a third position. Repeat steps 10–13.

CURVED MIRRORS

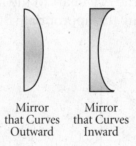

Mirror that Curves Outward   Mirror that Curves Inward

16. Obtain a curved mirror. Use one mirror support to hold the mirror upright on the bench. Place the mirror so that you are facing the side that curves outward (the convex side).

17. Place an object at various distances from the mirror. Look at the image of the object in the mirror.

18. Observe and record in your notebook how the image appears. Include the object's position (close to the mirror or far from the mirror), the size of the image (enlarged or small), and the orientation of the image (upright or upside down).

19. Turn the mirror around so that you are facing the side that curves inward (the concave side).

20. Place an object at various distances from the mirror. Look at the image of the object in the mirror.

21. Observe and record in your notebook how the image appears. Include the object's position (close to the mirror or far from the mirror), the size of the image (enlarged or small), and the orientation of the image (upright or upside down).

Mirror Images (cont.)

Analysis
VIRTUAL IMAGES

1. Describing events Describe the image of the reverse eye chart you saw on the surface of the mirror. Compare it with the appearance of the normal eye chart.

_____

_____

_____

2. Examining data What distance did you measure between the mirror and the reverse eye chart?

_____

_____

3. Examining data What distance did you measure between the starting point and the eye chart on the wall?

_____

_____

_____

FLAT MIRRORS

4. Organizing data In your notebook, draw the experimental setup as viewed from above. Include the lines and angles for each trial.

_____

_____

_____

CURVED MIRRORS

5. Describing events How did the image appear when the object was in front of the convex mirror?

_____

_____

_____

Mirror Images (cont.)

6. Describing events How did the image appear when the object was close to the concave mirror?

_____

_____

_____

7. Describing events How did the image appear when the object was far away from the concave mirror?

_____

_____

_____

Conclusions

1. Drawing conclusions Compare your answers to Analysis questions 2 and 3 above. What is the relationship between the distances?

_____

_____

_____

2. Drawing conclusions Compare the two angles measured in step 13 for each position. What is the relationship between the angles?

_____

_____

_____

3. Classifying Which kinds of mirrors used in this experiment are capable of producing images that are reversed left to right?

_____

_____

_____

4. Classifying Which kinds of mirrors used in this experiment are capable of producing images that are inverted (upside down)?

_____

_____

_____

Mirror Images (cont.)

Extensions

1. Research and communications Many side mirrors on automobiles have a warning label, "Objects in mirror are closer than they appear." What kind of mirror do you think these are? What might be the advantage of using this kind of mirror for this application? Research these types of mirrors on the Internet or in a library to check your answers.

2. Building models If you were designing a house of mirrors for a carnival, what kind of mirror would you use to create an upside down image? Would the person looking in the mirror have to be close to the mirror or far away from it to see his or her image upside down?

Skills Practice Lab

Converting Wind Energy into Electricity

Introduction

Electricity is a form of energy. Electricity can come from many different sources. The electricity in your house comes from a power plant, which may generate electricity by harnessing energy from burning coal or natural gas, nuclear reactions, flowing water, or wind. The energy from any one of these sources can be converted into electrical energy, which can in turn be converted into other forms of energy. For example, an electric stove or a toaster turns electrical energy into heat that can be used for cooking. A lamp turns electrical energy into light.

In this lab, you will build a device to convert the energy in wind into electrical energy. The electrical energy will then be converted into light as the electricity powers a light-emitting diode (LED).

OBJECTIVES

Construct a device that uses wind energy to generate electricity.

Use the device to power a light-emitting diode (LED).

Describe how changing various parts of the system affects the generation of electricity.

MATERIALS

alligator clips and wire (2)
ALNICO cylindrical cow magnet hole
   punch
blow dryer, 1500 W, 60 Hz
LED, 2.0 V
metric ruler
modeling clay
pencil, round and unsharpened

pipe cleaners (2)
poster board, 10 cm × 10 cm
PVC wire, 100 ft coil
scissors
shoe box
thumbtack
transparent tape
voltmeter or multimeter

Safety

- Do not place electrical cords in walking areas or let cords hang over a table edge in a way that could cause equipment to fall if the cord is accidentally pulled.

- Be sure that equipment is in the "off" position before plugging it in.

- Be sure to turn off and unplug electrical equipment when finished.

- If you are unsure of whether an object is hot, do not touch it.

- Use knives and other sharp instruments with extreme care.

Converting Wind Energy into Electricity (cont.)

- Never cut objects while holding them in your hands. Place objects on a suitable work surface for cutting.

Procedure
BUILDING A WINDMILL

1. Use the ruler as a straight edge to draw two diagonal lines on the poster board. Each line should go from one corner to the opposite corner. Punch a hole in the center of the square (where the lines cross). Make a 5 cm cut from each corner of the square toward the center (along the lines).

2. Slide the eraser end of the pencil through the hole in the center of the square so that the eraser end of the pencil sticks out about 3 cm. Carefully bend one corner of the square toward the eraser. Be careful not to make a fold in the poster board. Then, work your way around the square by bending alternating corners in to the eraser until the poster board is in a pinwheel shape, as shown in **Figure 1.** Use a thumbtack to pin the four folded corners to the eraser. This step can be tricky, so be patient and be careful not to poke yourself with the tack. When you are finished, you should have a turbine on a shaft.

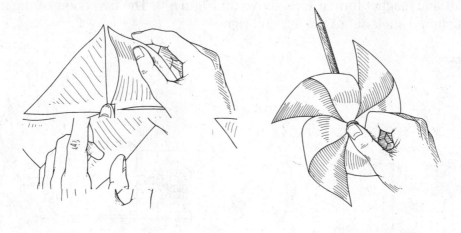

FIGURE 1

3. Push the back end of the turbine away from the eraser to create a lot of space in the turbine to catch wind. Securely tape the back end of the turbine to the pencil so that the shaft will turn when the turbine turns.

4. Place the box on the table so that the opening faces up. Carefully punch a hole in the long side of the shoebox, at least 8 cm above the table. Punch another hole on the other long side of the box so that the holes are opposite one another.

5. Use tape to attach the box to the table, or place a weight inside the box. Slide the shaft through the two holes so that the turbine is near the box and the other end of the shaft sticks out the other side of the box.

| Converting Wind Energy into Electricity (cont.)

6. Wrap pipe cleaners around the pencil where it exits the box on both sides. Tape the pipe cleaners to the pencil to prevent the shaft from sliding when wind blows on the turbine. Blow on the turbine to make sure that the turbine and shaft turn freely.

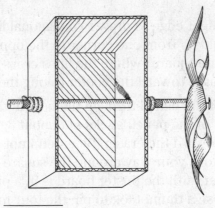

FIGURE 2

7. Tape the bar magnet to the end of the shaft opposite the turbine so that the shaft and magnet form a **T**, as shown in **Figure 3**. The two poles of the magnet should stick out to the left and right.

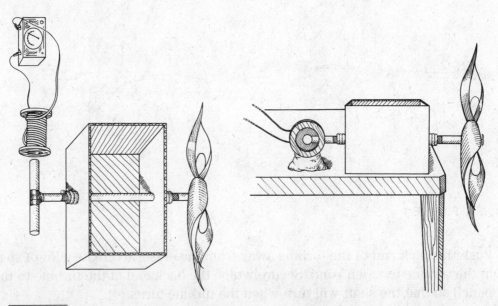

FIGURE 3

8. Place the coiled wire beside the magnet. Point the magnet's pole to the center of the coil. Place the coil on a base of modeling clay to hold the coil in position. Use the modeling clay to adjust the height of the coil so that the center of the coil is the same height as the magnet when it points to the coil. (The center of the coil should be at the same height as the shaft, about 8 cm above the table.)

GENERATING ELECTRICITY

9. Find the two loose ends of the wire. One wire is inside the coil, and the other wire is outside the coil. Tape the outside of the coil so that it does not unravel. Connect each loose wire end to a separate alligator clip and wire assembly. Clip the loose end of each alligator clip and wire assembly to a different voltmeter terminal.

10. Direct the blow dryer on the turbine. For best results, hold the blow dryer at an angle, as shown in **Figure 4.** Do not point directly at the center of the turbine but at one of the blades from the side. Be careful not to touch the turbine with the blow dryer.

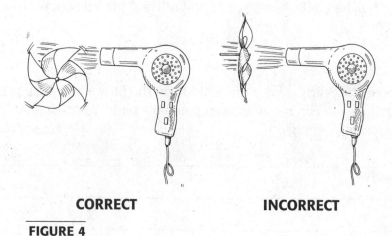

CORRECT **INCORRECT**

FIGURE 4

11. For the windmill to work properly, the turbine and shaft must not wobble when they spin. You must also have a good connection between the wires, and the wire coil must be very close to the magnet. If the magnet or shaft wobbles as it spins, adjust the magnet so that it is centered on the end of the shaft. Also, adjust the pipe cleaners around the pencil; a slight movement can change how much the magnet wobbles. Be patient; the alignment must be exact for the windmill to produce electrical energy.

12. Once the turbine seems to be balanced and the coil seems to be positioned properly, direct the blow dryer on the turbine again and read the voltage on the voltmeter.

13. If the voltmeter reads less than 1 V, turn off the blow dryer. Check the connections to the coil and to the voltmeter. Make sure that there are no breaks in the insulation on the wire. Adjust the wire spool and the spinning magnet so that they are as close together as possible without touching. Direct the blow dryer on the turbine again. If the voltage is still less than 1 V, move the coil to slightly different positions while the turbine is spinning to try to get a higher voltage. Be careful not to let the coil touch the magnet while the magnet is spinning.

| **Converting Wind Energy into Electricity (cont.)**

USING ELECTRICITY TO LIGHT AN LED

14. Once the voltmeter consistently reads over 1 V, turn off the blow dryer. Disconnect the voltmeter, and clip the alligator clips and wire assembly to each wire of an LED.

15. Direct the blow dryer on the turbine again. You should see the LED glow.

16. Move the blow dryer so that it points to the turbine at different angles. Observe how the brightness of the LED is affected.

17. While the LED is glowing, move the coil around so that the shaft points straight toward the center of the shaft. Observe how the brightness of the LED is affected.

18. When you are finished, take apart your windmill and put all supplies back in their proper place.

ANALYSIS

1. Describing events Starting with the wind from the blow dryer, list as many forms of energy as you can in the system you have built. Your list should have at least three forms of energy.

_____

_____

_____

_____

2. Recognizing patterns What happened to the brightness of the LED when you moved the blow dryer away from the position shown in Figure 4?

_____

_____

_____

_____

3. Explaining events Why did moving the blow dryer change the brightness of the LED?

_____

_____

_____

_____

Converting Wind Energy into Electricity (cont.)

4. Describing events What happened to the brightness of the LED when you moved the coil in step 17?

_____

_____

_____

_____

CONCLUSIONS

1. Drawing conclusions Based on your observations, in what position should the coil be relative to the spinning magnet in order to produce the most electricity?

_____

_____

_____

_____

_____

2. Making predictions Would the LED still glow if you turned the magnet so that it pointed straight out from the shaft?

_____

_____

_____

_____

3. Drawing conclusions Based on your observations, in what position should the blow dryer be relative to the turbine in order to produce the most electricity?

_____

_____

_____

_____

Converting Wind Energy into Electricity (cont.)

EXTENSIONS

1. **Building models** In a real-world situation, the wind blowing on a windmill does not always come from the same direction. How could you modify your windmill design to solve this problem?

   _____

   _____

   _____

   _____

2. **Building models** In a real-world situation, the wind does not always blow enough to turn a windmill. What could be added to a windmill system to ensure a steady supply of electricity when the wind stops blowing (at least for a while)?

   _____

   _____

   _____

   _____

Name _____ Class _____ Date _____

Constructing and Using a Compass

Introduction

Like a bar magnet, Earth has a magnetic field, with a magnetic north pole and a magnetic south pole. A compass takes advantage of Earth's magnetic field to orient a needle along the north-south direction. In a commercial compass, the north-pointing end of a compass needle usually has a colored arrow on it, so you can tell north from south. If you place a bar magnet near a compass, the north-pointing end of the needle will actually point toward the south pole of the magnet. This concept can be confusing—just remember that the North Pole of the Earth is like the south pole of a magnet.

In this activity, you will use ordinary objects—a paper clip, a film-canister lid, and a container of water—to construct a working compass. You will then predict and test how different objects will affect the compass. If your teacher has prepared an orienteering course, you may use your compass to navigate the course.

OBJECTIVES

Construct a compass by using simple materials.

Predict how various objects will affect the compass.

Use the compass to navigate a short orienteering course.

MATERIALS

aluminum can	marker, permanent
bar magnets, strong (2)	paper clip, large
film-canister lid	plastic cup
index card with directions	water
iron nail	
margarine tub or other small, shallow container	

SAFETY

- Use knives and other sharp instruments with extreme care.

Procedure
CONSTRUCT A COMPASS

1. Stroke the paper clip 50 times with one end of a magnet. Stroke in one direction only. Stroking in two directions will not magnetize the paper clip. Be careful not to scrape or puncture yourself with the paper clip.

71 Laboratory Manual

| Constructing and Using a Compass (cont.)

2. Fill the container to the halfway point with water. Gently place a film-canister lid upside down on the surface of the water in the center of the container.

3. Predict what will happen when you set the paper clip on the lid.

4. Test your prediction by gently setting the paper clip on the film-canister lid and observing what happens.

5. Lift the paper clip, and place it back on the lid in a different direction. Repeat this step several times.

6. If all has gone well, you have just made a compass. To determine which end of the paper clip points north, put the south end of the bar magnet about 10 cm from the compass. Use the permanent marker to mark the end of the paper clip that points to the south end of the bar magnet.

7. Carefully remove the paper clip and the canister lid from the water. From this point on, the paper clip will be referred to as the needle of the compass.

8. Use a permanent marker to label all four compass points (N, S, E, and W) on the face of the canister lid.

9. Float the lid in the water. Put the compass needle back on the lid so that the needle points north. You are now ready to use your compass!

Hint: The needle should stay magnetized throughout the activity. However, if it is dropped, it may need to be magnetized again, as in step 1.

TESTING YOUR COMPASS

10. Before you use the compass as a tool, you should discover what might interfere with its operation. Predict how each object in the table below might affect the operation of the compass. Record your predictions in a table like the one below.

11. Test your predictions by placing each object 5 cm from the compass (but not in the north or south directions). Record your observations in your data table.

COMPASS RESPONSE DATA

Object	Make a prediction	Conduct an experiment	Make observations
	Will the compass needle move?	Place the object 5 cm from the compass	Did the needle move?
Aluminum can			
Iron nail			
Magnet			
Plastic cup			

Constructing and Using a Compass (cont.)

USING YOUR COMPASS TO NAVIGATE A COURSE (OPTIONAL)

This part of the Procedure is to be done outdoors in an area prepared by your teacher. In this activity, you will use the compass you have constructed to navigate a course to find objects laid out by your teacher. Your teacher should give you an index card containing directions for you to follow on the course.

12. In many cases, it is impractical to measure distance with a meterstick or a tape measure. Instead, you can measure distance by paces. Your teacher has marked off a 10 m distance. Count your steps as you walk the 10 m. Your steps should be regular and consistent. On the back of the index card, write down the number of steps you walked.

13. Repeat step fifteen twice, and write down the number of steps you walked. In all three cases and in navigating the course, the same person should do the pacing and should take care to keep each step the same length.

14. Calculate your average number of steps in 10 m by adding the total number of steps you took in all three trials and dividing by 3. Write this number down on the back of the index card.

15. Move to the starting point indicated in your directions.

16. Hold the compass, and observe which direction is north.

17. Read the first step of your directions, and determine the direction to walk. Face that direction.

18. Calculate the number of steps required to walk the specified distance. Walk that number of steps.

19. When you reach the destination, you will find an object and an index card containing the name of the object. Record the name of the object in the appropriate place on your index card.

20. Repeat steps 17–19 for each direction on your card until you have listed all five objects.

21. Take your index card to your teacher. If you have successfully completed the course, continue to the Analysis. If you had any difficulty, repeat the course until you are successful.

ANALYSIS

1. **Explaining events** Describe what you observed when you placed the paper clip on the film-canister lid in step 4. Why do you think this happened?

_____

_____

_____

_____

2. Explaining events How did each of the objects affect the compass in step 11? Explain your results.

_____

_____

_____

_____

CONCLUSIONS

1. Evaluating results Was your prediction in step 3 of the Procedure correct? Why or why not?

_____

_____

_____

_____

2. Making predictions What would have happened if you had placed the paper clip on the film-canister lid before you stroked the paper clip with a magnet?

_____

_____

_____

EXTENSIONS

1. Building models Imagine that you are lost in the woods without a compass. You have a map of the area, but you do not know which direction is north. It is only a couple of hours before sunset, and you need to get back to your camp. You have with you several topographic maps held together with a paper clip, a bottle of water, a camera and several canisters of film, a plastic storage container of trail mix, and, luckily, a magnet on your key chain. Describe how you could use these materials to build a compass.

_____

_____

_____

_____

_____

_____

Constructing a Radio Receiver

Introduction

You have probably listened to radios many times in your life. Modern radios are complicated electronic devices. However, radios do not have to be so complicated. The basic parts of any radio include a diode, a capacitor, an antenna, a ground wire, and an earphone (or a speaker and amplifier in a large radio). In this experiment, you will examine each of these components one at a time as you construct a working radio receiver.

OBJECTIVES

Construct a radio receiver from simple materials.

Use the receiver to detect radio signals.

Explain how different parts of the receiver contribute to the operation of the receiver.

Compare the receiver to a fully functioning radio.

MATERIALS

aluminum foil	ground wire
antenna	insulated wire, 2 m long
cardboard, 20 cm × 30 cm	paper clips (3)
cardboard tubes (2)	scissors
connecting wires, 30 cm long (7)	sheet of paper
diode	tape
earphone	

SAFETY

- Secure loose clothing and remove dangling jewelry. Don't wear open-toed shoes or sandals in the lab.

- Do not place electrical cords in walking areas or let cords hang over a table edge in a way that could cause equipment to fall if the cord is accidentally pulled.

- Use knives and other sharp instruments with extreme care.

- Never cut objects while holding them in your hands. Place objects on a suitable work surface for cutting.

Procedure

MAKING AN INDUCTOR AND A CAPACITOR

1. Examine the diode. An electric current in a diode is in only one direction. On a sheet of paper or in your lab notebook, describe the appearance of the diode.

2. An inductor controls the amount of electric current by changing the amount of resistance in a wire. Make an inductor by wrapping a 2 m piece of insulated wire around a cardboard tube approximately 100 times. Leave a few centimeters of the cardboard exposed on each end. Wind the wire so that each turn of the coil touches but does not overlap the previous turn. Leave about 25 cm of wire loose on each end of the coil. The turns of the coil should end up in a neat and orderly row that extends across most of the length of the tube, as shown in **Figure 1.** Use tape to secure the coil to the tube. Set your inductor aside for now.

FIGURE 1

3. A capacitor stores electrical energy when a current is applied. A variable capacitor is a capacitor in which the amount of stored energy can change. You will make a variable capacitor in steps 4–6.

4. Cut a piece of aluminum foil to go around half the length of another cardboard tube. Keep the foil as wrinkle free as possible as you wrap it around the tube. Tape the foil to itself to keep it wrapped around the tube, and then tape the top and bottom of the foil to the tube to keep the foil from sliding.

5. Use the sheet of paper and tape to make a sliding cover over the foil on the tube. The paper should completely cover the foil on the tube with about 1 cm extra. Wrap the paper so that it fits snugly around the foil, but leave it loose enough that it can slide up and down over the foil.

6. Cut another piece of aluminum foil to wrap completely around the paper. Leave approximately 1 cm of paper showing beyond each end of the foil. Keep the foil as wrinkle free as possible as you wrap it around the tube. Do not make the foil so tight that the paper can no longer slide over the bottom layer of foil. Tape the foil to itself to keep it wrapped around the paper, and then tape the top and bottom of the foil to the paper.

7. Stand the tube on end so that the inner layer of foil is on the bottom half. Slide the paper and foil sleeve up so that it covers the top half of the tube.

Constructing a Radio Receiver *(cont.)*

This tube with the two layers of foil can now serve as a variable capacitor. The amount of stored energy depends on how much the two layers of foil overlap. Therefore, the capacitor stores more energy the farther down the paper and foil sleeve is on the tube. The completed capacitor tube is shown as part of **Figure 2.**

ASSEMBLING THE RECEIVER

8. Use tape to attach a 30 cm connecting wire to the inner foil layer at the bottom of the variable capacitor tube. Also use tape to attach another 30 cm connecting wire to the foil at the bottom of the foil and paper sleeve. Make sure that the metal ends of each wire make good contact with the foil. **Figure 2** shows these wires attached to the capacitor.

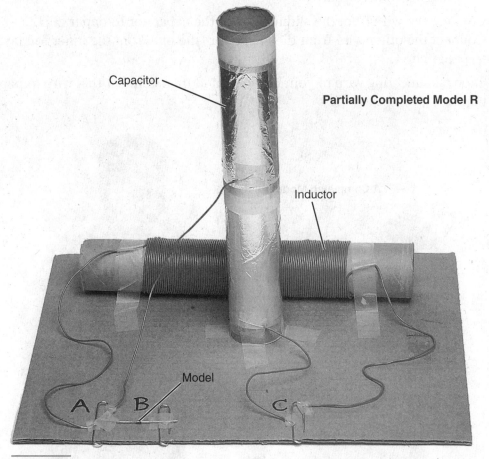

Capacitor

Partially Completed Model R

Inductor

Model

A   B   C

FIGURE 2

9. Hook three paper clips on one edge of a 20 cm × 30 cm piece of cardboard, as shown in **Figure 2.** Label one paper clip "A," another one "B," and the third one "C."

10. Lay the inductor (the tube that you made earlier with the coiled wire) on the cardboard, and tape it to the cardboard, as shown in **Figure 2.**

11. Place the capacitor, still upright in the same orientation, next to the inductor (but not touching it), and tape the bottom of the tube to the cardboard. The tape may cover part of the inner layer of foil, but be careful not to tape the paper and foil sleeve—the sleeve must be free to slide.

12. Use tape to connect the diode to paper clips A and B. The cathode should be closest to paper clip A. (The cathode side of the diode is the side with the dark band; if you are unsure about which direction to place the diode, ask your teacher.) Make sure that all connections have good metal-to-metal contact.

13. Connect the wire from one end of the inductor to paper clip A and the wire from the other end to paper clip C. Use tape to hold the wires in place, and make sure that there is good metal-to-metal contact. You may need to strip some of the insulation off the wire near the tip to get good contact. If so, ask your teacher to help you.

14. Connect the wire from the sliding part of the capacitor to paper clip A. Connect the other wire from the capacitor (the one from the inner foil layer) to paper clip C.

15. Tape a connecting wire to your antenna, and then connect this wire to paper clip A.

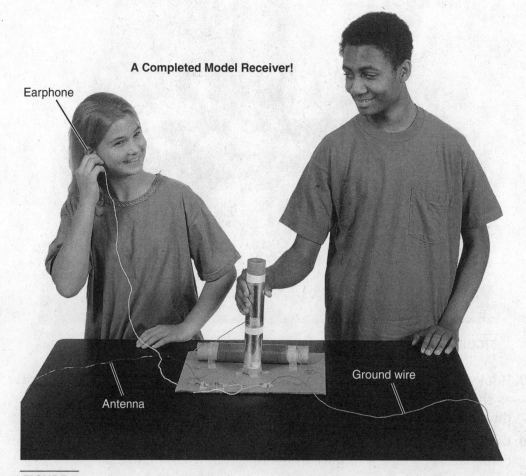

A Completed Model Receiver!

Earphone

Ground wire

Antenna

FIGURE 3

Constructing a Radio Receiver (cont.)

16. Use tape to connect one end of the ground wire to paper clip C. The other end of the wire should be connected to a "ground." This "ground" may be a plumbing fixture or some other point specified by your teacher.

17. Connect one wire from the earphone to paper clip B and the other wire to paper clip C, as shown in **Figure 3.**

USING YOUR RECEIVER

18. Your radio receiver is now complete. It should look similar to the receiver in **Figure 3.** The antenna will pick up radio waves in the air. These waves may be transmitted by a radio station, or they may be "noise" or "static," which are radio waves coming from particles in the atmosphere or even from space. The earphone will allow you to hear the signals (or the noise) corresponding to the radio waves that the antenna picks up.

19. It may take some work to get an audible signal through your receiver. If you do not hear any sound at all, you may need to troubleshoot by checking all of the parts of your receiver one by one. Make sure that there is good contact between metal connections where there is supposed to be contact, and make sure that there is no contact where there is not supposed to be contact (such as between the two pieces of foil in the capacitor or between the metal cores of the wires in the inductor coil). Make sure that the diode is facing the right way and that the ground wire is grounded. Remember that even audible static is evidence that your receiver is working properly. If you have tried troubleshooting and still cannot hear any sound, ask your teacher for help. Be sure to let everyone in the group have a chance to listen to and experiment with the receiver.

ANALYSIS

1. Describing events Describe the process you used to obtain a signal from the receiver.

_____

_____

_____

2. Describing events Every capacitor consists of two conductors separated by an insulator. List the two conductors and the insulator in your capacitor.

_____

_____

3. Explaining events The amount of energy stored in a capacitor depends in part on the overlapping area of the two plates. As you moved the paper and foil sleeve downward, did the energy storage of the capacitor increase or did it decrease? Explain.

_____

_____

_____

4. Explaining events One purpose of an inductor in a radio receiver is to increase electrical resistance. The resistance in a wire increases as the length of the wire increases. Why is the resistance in the coil of wire in an inductor greater than the resistance in a straight wire that spans the same length as the inductor?

_____

_____

_____

5. Explaining events What kind of wave are the radio waves received by your antenna? What kind of wave passes the signal from the earphone to your ear?

_____

_____

_____

CONCLUSIONS

1. Making predictions A diode allows current in only one direction. What would happen to your receiver if you turned your diode around so that it faced the wrong way?

_____

_____

_____

2. Drawing conclusions Are the radio waves picked up by your receiver analog signals or digital signals?

_____

_____

Constructing a Radio Receiver (cont.)

3. Evaluating models List at least two ways that your receiver is similar to a fully functional radio. List at least two ways that your receiver differs from a fully functional radio.

_____

_____

_____

_____

_____

_____

EXTENSIONS

1. Research and communications Research the design of radio transmitters. Write a paragraph that describes some of the materials that you would need in order to build a transmitter to transmit signals that you could pick up with your receiver.

Name _____ Class _____ Date _____

Explaining the Motion of Mars

Introduction

If you watch the planet Mars every night for several months, you will notice that it appears to "wander" among the stars. While the stars remain in fixed positions relative to each other, the planets appear to move independently of the stars. Mars first travels to the left, then it goes back to the right a little, and finally it reverses direction and travels again to the left. For this reason, the ancient Greeks called the planets "wanderers."

The ancient Greeks believed that the planets and the sun revolved around Earth and that Earth was the center of the solar system. The Greek mathematician Ptolemy came up with a complicated system of "epicycles" that explained why the planets appeared to wander as they orbited Earth (or so he thought).

Ptolemy's model of the solar system was believed for centuries. In the early 1500s, Nicolaus Copernicus proposed that the sun was the center of the solar system and that all the planets, including Earth, revolved around the sun. Copernicus's new model was convincing and was eventually widely accepted because it explained the wandering motion of the planets much better than the Earth-centered model did.

In this lab, you will make your own model of part of the solar system to find out how Copernicus's model of the solar system explained the zigzag motion of the planets.

OBJECTIVES

Illustrate the apparent motion of Mars as seen from Earth against the background of stars.

Explain why Mars moves in a zigzag pattern.

Predict the apparent motion of Jupiter as seen from Earth.

MATERIALS

colored pencils   metric ruler
drawing compass   white paper

Safety

- Use knives and other sharp instruments with extreme care.

Procedure

ASK A QUESTION

1. Why do the planets appear to move back and forth against the background of stars in Earth's night sky? Propose a hypothesis to explain this motion, and write down your hypothesis.

CONDUCT AN EXPERIMENT

2. Mark a point on the center of a piece of paper. Label the point "Sun."

3. Place the point of the compass on the point that represents the sun. Use the compass to draw a circle with a diameter of 9 cm. This circle will represent the orbit of Earth around the sun. (Note: The orbits of the planets are slightly elliptical, but circles will work for this activity.)

4. Using the same center point, draw a circle with a diameter of 12 cm. This circle will represent the orbit of Mars.

5. Using a blue pencil, draw three parallel lines in a diagonal across one end of your paper, as shown in **Figure 1.** These lines will help you plot the path Mars appears to travel in Earth's night sky. Turn your paper so that the diagonal lines are at the top of the page.

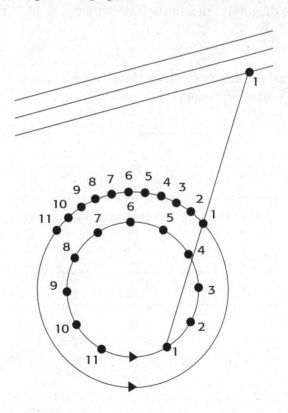

FIGURE 1

6. Place 11 dots on the orbit of Earth and number them 1 through 11 as shown in **Figure 1.** These dots represent Earth's position from month to month.

Explaining the Motion of Mars (cont.)

7. Now place 11 dots along the top of the orbit of Mars. Number the dots as shown from month to month. These dots represent the position of Mars in the figure. Notice that Mars travels slower than Earth.

8. Use a green line to connect the first dot on Earth's orbit to the first dot on Mars's orbit, and extend the line until it meets the lowest of the three diagonal blue lines at the top of the paper. Use the ruler as a straight edge to make sure your line is straight. Place a green dot where this green line meets the lowest blue diagonal line, and label the green dot "1."

9. Now connect the second dot on Earth's orbit to the second dot on Mars's orbit, and extend the line until it meets the lowest blue diagonal. Place a green dot where this line meets the lowest blue diagonal line, and label this dot "2."

10. Continue drawing green lines from Earth's orbit to Mars's orbit and up to the blue diagonal lines. Pay attention to the pattern of dots that you are adding to the diagonal lines. When the direction of the dots changes, extend the green line to the middle blue diagonal line, and add the dots to that line instead. If the direction of the dots changes again, extend your green lines to the top-most blue diagonal line, and add the dots to that line instead.

11. When you are finished adding green lines, draw a red line to connect all of the dots on the blue diagonal lines in the order that you drew the dots.

ANALYSIS

1. **Classifying** What do the green lines that connect points along Earth's orbit and along Mars's orbit represent?

_____

_____

_____

2. **Classifying** What does the red line that connects the dots along the diagonal lines represent?

_____

_____

_____

3. Describing events Describe the path of Mars as it would be seen from Earth over the course of the year represented in this activity. Assume Earth is at position 1 in January. In your description, name the months in which Mars would appear to change directions.

_____

_____

_____

_____

CONCLUSIONS

1. Drawing conclusions State in your own words why Mars appears to move in a zigzag motion when viewed from Earth against the background of stars.

_____

_____

_____

_____

2. Making predictions Do you think the planet Jupiter would appear to move in a zigzag motion when viewed from Earth against the background of stars? Explain why or why not.

_____

_____

_____

_____

EXTENSIONS

1. Building models Unlike the orbits of Mars and Jupiter, the orbit of Venus around the Sun is inside Earth's orbit. Repeat this experiment, but let the outer circle represent Earth's orbit and the inner circle represent the orbit of Venus. Draw lines from each position of Earth through the corresponding positions of Venus, and extend the lines to the bottom of the page. Would Venus appear to move in a zigzag motion? How would Venus appear to move *relative to the Sun* during the year?

Name _____ Class _____ Date _____

Classifying Rocks

Introduction

There are many different types of igneous, sedimentary, and metamorphic rocks. Therefore, one must to know important distinguishing features of the rocks in order to classify the rocks. The classification of rocks is generally based on their mode or origin, their mineral composition, and the size and arrangement (or texture) of their minerals. The many types of igneous rocks differ in the minerals that they contain and the sizes of their crystalline mineral grains. Igneous rocks composed of large mineral grains have a coarse-grained texture. Some igneous rocs have small mineral grains that cannot be seen with the unaided eye. These types of rocks have a fine-grained texture.

Many sedimentary rocks are made of fragments of other rocks compressed and cemented together. Some sedimentary rocks have a wide range of sediment sizes, while others may have only one size. Other common features of sedimentary rocks include parallel layers, ripple marks, cross-bedding, and the presence of fossils.

Metamorphic rocks often look similar to igneous rocks, but they may have bands of minerals. Metamorphic rocks with a foliated texture have minerals arranged in bands. Metamorphic rocks that do not have bands of minerals are nonfoliated.

In this investigation, you will use a rock identification table to identify various rock samples.

OBJECTIVES

Observe several properties of rock samples, including color, texture, composition, and whether samples react with acid.

Identify the class and name of each sample of rock by using observations and a rock identification table.

MATERIALS

dilute hydrochloric acid, 10%   rock samples
hand lens   safety goggles
medicine dropper

SAFETY

- If a chemical gets on your skin or clothing or in your eyes, rinse it immediately, and alert your instructor.

- If a chemical is spilled on the floor or lab bench, alert your instructor, but do not clean it up yourself unless your teacher says it is OK to do so.

Classifying Rocks (cont.)

• Wear safety goggles when working around chemicals, acids, bases, flames, or heating devices.

• If any substance gets in your eyes, notify your instructor immediately and flush your eyes with running water for at least 15 minutes.

Procedure

1. List in your data table the numbers of the rock samples that your teacher gave you.

DATA TABLE

Specimen	Description of properties	Rock class	Rock name

2. Using a hand lens, study the rock samples. Look for characteristics such as the shape, size, and arrangement of the mineral crystals. For each sample, list in your table the distinguishing features that you observe.

3. Certain rocks react with acid, which indicates that they are composed of calcite. If a rock contains calcite, it will bubble, releasing carbon dioxide. Using a medicine dropper and 10% dilute hydrochloric acid, test your samples for their reactions. **CAUTION:** *Wear goggles when working with hydrochloric acid.*

4. Refer to the rock identification table. Compare the properties for each rock sample that you listed with the properties listed in the identification table. If you are unable to identify certain rocks, examine these rock samples again.

5. Complete your table by identifying the class of rocks—igneous, sedimentary, or metamorphic—that each rock sample belongs to, and then name the rock.

| Classifying Rocks (cont.)

Description	Rock class	Rock name
Coarse-grained; mostly light in color—shades of pink, gray, and white are common	igneous	granite
Coarse-grained; mostly dark in color; much heavier than granite or diorite	igneous	gabbro
Fine-grained; dark in color; often rings like a bell when struck with a hammer	igneous	basalt
Light to dark in color; many holes—spongy appearance; light in weight; may float in water	igneous	pumice
Light to dark in color; glassy luster—sometimes translucent; conchoidal features	igneous	obsidian
Coarse-grained; foliated; layers of different minerals often give a banded appearance	metamorphic	gneiss
Coarse-grained; foliated; quartz abundant; commonly contains garnet; flaky minerals	metamorphic	schist
Fine-grained; foliated; cleaves into thin, flat plates	metamorphic	slate
Coarse-grained; nonfoliated; reacts with acid, effervesces	metamorphic	marble
Fine-grained; soft and porous; normally white or buff in color	sedimentary	chalk
Coarse-grained, over 2 mm; rounded pebbles; some sorting—clay and sand can be seen	sedimentary	conglomerate
Medium-grained, 1/16 mm to 2 mm; mostly quartz fragments—surface feels sandy	sedimentary	sandstone
Microscopic grains; clay composition; smooth surface—hardened mud appearance	sedimentary	shale
Coarse- to medium-grained; well-preserved fossils are common; soft—can be scratched with a knife; occurs in many colors but usually white-gray; reacts with acid	sedimentary	crystalline limestone
Coarse- to fine-grained; cube-shaped crystals; normally colorless; does not react with acid	sedimentary	halite

Classifying Rocks (cont.)

ANALYSIS

1. Identifying/recognizing patterns What properties were most useful in identifying each rock sample?

_____

_____

_____

2. Classifying Were any samples difficult to identify? Explain.

_____

_____

_____

3. Classifying Were any characteristics common to all of the rock samples?

_____

_____

_____

CONCLUSIONS

1. Drawing conclusions How can you distinguish between a sedimentary rock and a foliated metamorphic rock when both types of rock have observable layering?

_____

_____

_____

Classifying Rocks (cont.)

2. **Applying conclusions** Name properties that distinguish between the rocks in the following pairs.

   a. granite and limestone

_____

_____

_____

   b. obsidian and sandstone

_____

_____

_____

   c. pumice and slate

_____

_____

_____

   d. conglomerate and gneiss

_____

_____

_____

EXTENSIONS

1. **Research and communications** Collect a variety of rocks in your area. Use the rock identification table to see how many rocks you can classify. How many rocks are igneous? How many are sedimentary rocks? How many are metamorphic rocks? After you identify the class of each rock, try to name the rock.

Building a Cup Anemometer

Introduction

An anemometer is a device that measures wind speed. Anemometers are used in weather stations and also on airplanes and ships to measure speed relative to the air. Anemometers have been around for a long time. Drawings by Leonardo da Vinci show his design of a deflection anemometer, in which the wind lifts a plate at an angle corresponding to the wind speed. Many modern anemometers use cups that catch the wind, which causes an axis to rotate at a speed corresponding to the wind speed. This type of anemometer was first designed in 1850 by Dr. Thomas Robinson. In this lab, you will build a simple cup anemometer, and you will use it to measure approximate wind speed.

OBJECTIVES

Construct a cup-style anemometer by using simple materials.

Use an anemometer to measure wind speed.

Compare wind speed measurements with predicted values and with values obtained by others.

Evaluate the design of the anemometer, and suggest improvements.

MATERIALS

hole punch	pencil, sharp with an eraser
marker, colored	plastic straws, straight (2)
masking tape	scissors
metric ruler	stapler
modeling clay	stopwatch
paper cups, small (5)	straight pin

SAFETY

• Secure loose clothing and remove dangling jewelry. Don't wear open-toed shoes or sandals in the lab.

• Use knives and other sharp instruments with extreme care.

• Never cut objects while holding them in your hands. Place objects on a suitable work surface for cutting.

Procedure

BUILDING THE ANEMOMETER

1. Cut off the rolled edges of four of the five paper cups. This will make the cups lighter so that they can spin more easily.

2. Make four evenly spaced markings around the outside of the cup that still has its rolled edge. The marks should be 1 cm below the rim of the cup and should go all the way around the cup so that each mark has another mark directly opposite it on the other side of the cup.

3. Use the hole punch to punch a hole at each of the marks on the cup. Use the sharp pencil to carefully punch a hole in the center of the bottom of the cup.

4. Push a straw through two opposite holes in the sides of the cup. Push another straw through the other two opposite holes. The two straws should cross in the center of the cup.

5. Make a mark 3 cm from the bottom on each of the other four cups. At each mark, punch a hole in the cup by using the hole punch.

6. Use the colored marker to color the outside of one of the four cups. The color will allow you to pick out this cup when the cups are spinning.

7. Slide a cup onto one of the straws by pushing the end of the straw through the hole in the cup. Turn the cup so that the bottom of the cup faces to the right when the center cup is facing up.

8. Fold the end of the straw, and staple it to the inside of the cup directly across from the hole (the straw should stretch across the cup).

9. Repeat steps 7–8 for the three remaining cups.

10. Push the pin down through the two straws where they intersect in the middle of the center cup. Use caution because the pin is sharp. Push the eraser end of the pencil up through the bottom of the center cup. Push the pin as far as it will go into the eraser end of the pencil. Be careful not to push too hard—you do not want the pin to bend.

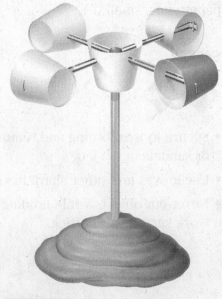

11. Push the sharpened end of the pencil into a solid base of modeling clay. The device should be able to stand upright even when a moderate wind is blowing. Your anemometer should now appear similar to the one shown in **Figure 1.**

12. Blow into the cups so that they spin. Adjust the pin if necessary so that the cups can spin freely without wobbling. Adjust the base if necessary to add stability.

FIGURE 1

Building a Cup Anemometer (cont.)

13. Mark a point on the base with masking tape. Label the tape "Starting point."

USING YOUR ANEMOMETER

14. Find a suitable area outside to place the anemometer on a level surface away from objects that would obstruct the wind, such as buildings or trees. If there is no wind at all or no appropriate place outside, your teacher may set up a fan in the lab.

15. Make a prediction about what the wind speed is at the location where you have placed your anemometer.

16. Hold the colored cup over the starting point in one hand while you or your partner holds the stopwatch. Release the colored cup, and start the stopwatch. As the cups spin, count the number of times the colored cup crosses the starting point in 10 s.

ANALYSIS

1. Describing events How many times did the colored cup cross the starting point in 10 s?

_____

_____

_____

2. Organizing data Divide your answer in question 1 by 10 to get the average number of revolutions in 1 s.

_____

_____

_____

3. Organizing data Measure the diameter of your anemometer (the distance from the outer edge of one cup to the outer edge of the opposite cup). Multiply this number by π (≈ 3.14) to get the circumference of your anemometer. Convert the circumference to meters.

_____

_____

_____

Building a Cup Anemometer (cont.)

4. **Organizing data** Multiply the circumference of the anemometer in meters by the number of revolutions per second that you calculated to get the wind speed in meters per second.

_____

_____

_____

5. **Organizing data** Convert the wind speed in meters per second to miles per hour. There are 1603 meters in a mile and 3600 seconds in an hour.

_____

_____

_____

CONCLUSIONS

1. **Evaluating results** Does the value you got for the wind speed seem reasonable? Is it close to what you predicted?

_____

_____

_____

_____

_____

2. **Evaluating results** Compare the wind speed you measured with the wind speeds measured by your classmates. If there are differences, identify at least two potential causes of the differences.

_____

_____

_____

_____

_____

Building a Cup Anemometer (cont.)

3. Evaluating models How could you modify your anemometer to make it more accurate or more reliable?

_____

_____

_____

_____

_____

EXTENSIONS

1. Designing experiments Use the anemometer to measure wind speed once a day for several days or once a week for several weeks. Check your values against the daily average values given by a local weather station. Are your values consistently higher or lower than the values given by the station? How could you explain this difference? Make a graph of the wind speeds versus time. Do you notice any patterns in the wind speed over time? Try to correlate the wind speeds with other weather events, such as the passage of warm fronts and cold fronts. Write a report that presents your data and summarizes the results.

Skills Practice Lab

Making Your Own Recycled Paper

Introduction

Paper has been made of many different materials throughout history. In the old days, Europeans often used animal skin to make paper. The Egyptians used grass. Paper in Asia is often made from rice. Most paper in the United States today is made from trees, which can take a toll on the environment. Your city or town may have a paper recycling program. However, even commercial recycling of paper uses precious energy resources, and the recycled paper is often supplemented with fresh wood pulp.

One easy and fun way that you can help the environment is to make your own recycled paper from paper scraps. You can even incorporate other fibers or natural dyes into your paper to make the paper more artistic. You can then use your homemade paper to write letters, send greeting cards, or keep a journal. In this activity, you will learn the basics of making your own paper. Once you know how, you can experiment to find unique ways to customize and decorate your paper.

OBJECTIVES

Create recycled paper from paper scraps.

Describe how paper fibers change throughout the papermaking process.

Explain how recycling paper can help the environment.

Evaluate this method of making recycled paper.

MATERIALS

cornstarch, 5 mL
cotton balls
cup, 500 mL
duct tape
embroidery hoop, 20 cm to 26 cm
   in diameter
long-handled spoon
nylon stocking, no runs

roasting pan, at least 10 cm deep
scissors
scrap paper
sponges (3 or 4)
towels (3 or 4)
water

SAFETY

• Secure loose clothing and remove dangling jewelry. Don't wear open-toed shoes or sandals in the lab.

• Always use caution when working with chemicals.

• Never mix chemicals unless specifically directed to do so.

- Wear an apron or lab coat to protect your clothing when working with chemicals.

- Follow instructions for proper disposal.

- If a spill gets on your clothing, rinse it off immediately with water for at least 5 minutes while notifying your instructor.

- Always wear protective gloves when handling chemicals.

- Do not eat any part of a plant or plant seed used in the lab.

- Wash hands thoroughly after handling any part of a plant.

- Use knives and other sharp instruments with extreme care

- Never cut objects while holding them in your hands. Place objects on a suitable work surface for cutting.

- Never use a double-edged razor in the lab.

Procedure

1. Separate the inner and outer parts of the embroidery hoop. Put the inner hoop into the stocking, being careful not to rip the stocking or cause a run. When the hoop is completely covered, replace the outer hoop and tighten the screw so that it fits snugly. The hoop is now a papermaker.

2. Place four pieces of duct tape along the edges of the hoop so that the center forms a rectangle, as shown in **Figure 1.**

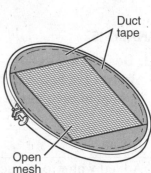

Duct tape

Open mesh

FIGURE 1

3. Add 5 mL of cornstarch to a 500 mL cup. Adding cornstarch to the paper will prevent ink from bleeding. Shred paper into small pieces (very roughly 1 inch square). Loosely pack the shreds into the cup until the cup is full.

4. Your teacher should have set up a central station for blending your paper scraps into a smooth pulp, called *slurry*. At the slurry station, carefully add 1 L of water to the blender. Add half of the contents of the cup to the blender, and securely place the lid on the blender. Your teacher will blend the mixture until it forms into slurry. Carefully pour the slurry into a pan that is at least 10 cm deep. Use the remainder of the cup's contents to repeat this process.

5. Your teacher should have set up another station that contains various items for decorating your paper. At the decoration station, choose various items to incorporate into your paper. Add them to your slurry, and blend the mixture with a spoon. Be careful not to add too much decorative material; such material could cause the paper to be too weak and to tear easily.

6. Separate the fibers of several cotton balls, and add them to the slurry to strengthen the paper. Stir the fibers into the slurry with a spoon until they are evenly distributed.

Making Your Own Recycled Paper (cont.)

7. Use tape to label a towel with your name. Unfold the towel, and place it beside the pan. Hold the papermaker tape-side up. Scoop the slurry into the papermaker, and let the slurry rest for a few minutes on the bottom of the pan. Fibers will settle on top of the papermaker.

8. Without tilting the papermaker, slowly lift it out of the pan. Let the water drain into the pan. When the papermaker stops dripping, carefully flip it onto the towel so that the new paper lies between the papermaker and the towel.

9. Gently press on the nylon with the sponge, and rub the sponge along the back of the mesh to absorb the water. Removing the excess water strengthens your paper and helps it dry quickly. When your sponge is full of water, wring the water into the pan. Repeat this process until you can no longer remove water from the paper.

10. Carefully lift the papermaker off your new paper sheet. Gently lift the towel with the paper on it, and move the towel and the paper to the drying area. (Hint: If the paper starts to rip, press the edges of the tear together with wet fingers.) Congratulations! You have made your own paper!

11. Repeat steps 6–10 for each group member.

12. Discard any unused pulp mixture into a compost pile or a container that your teacher has provided. Do not pour the slurry down the drain. Slurry clogs pipes.

ANALYSIS

1. **Describing events** The paper you have made is composed of dried plant fibers. Fill in the table below to describe what you think happened to the fibers in each of the steps that you followed.

Action	What happened to the fibers?
Added water to the scrap paper	
Blended the scrap paper to make slurry	
Lifted the papermaker from the pan	
Dried the paper	

| *Making Your Own Recycled Paper (cont.)*

2. Analyzing results How many pieces of paper did your group originally use to make the scraps that became your slurry? How many pieces of recycled paper did your group make?

_____

_____

_____

CONCLUSIONS

1. Drawing conclusions How does recycling paper help the environment?

_____

_____

_____

_____

2. Evaluating results How could this lab activity be modified to better conserve resources?

_____

_____

_____

_____

EXTENSIONS

1. Designing experiments If you have time and if your teacher has provided the materials, you can use natural materials to decorate your paper once it has dried. Your teacher should provide pigments made from flowers, fruit, vegetables, or spices. Be sure to wear a lab apron and disposable gloves while working with the pigments. You can use sponges to create textured patterns with the pigments on your paper. You can cut the sponges into different shapes to make stamps and then use the stamps to make repeated shapes on the paper. You can also use a feather dipped in pigment to write or draw on the paper. Let the pigments dry before moving or handling the paper.